T0224971

SpringerBriefs in Applied Sciences and Technology

Nanotheranostics

Series Editors

Subramanian Tamil Selvan, Institute of Materials Research & Engineering,
National University of Singapore, Singapore, Singapore

Karthikeyan Narayanan, Singapore, Singapore

Padmanabhan Parasuraman, Singapore, Singapore

Paulmurugan Ramasamy, School of Medicine, Stanford University, Palo Alto, CA,
USA

Indexed by SCOPUS. Nanotheranostics is a burgeoning field in recent years, which makes use of "nanotechnology" for diagnostics and therapy of different diseases. The recent advancement in the area of nanotechnology has enabled a new generation of different types of nanomaterials composed of either inorganic or polymer based nanoparticles to be useful for nanotheranostics applications. Some of the salient features of the nanotechnology towards medicine are cost reduction, reliable detection and diagnosis of diseases at an early stage for optimal treatment. The advent of nanotheranostics is expected to benefit the pharmaceutical and healthcare industries in the next 5-10 years. Nanotechnology holds a great potential to be explored as a multifunctional platform for a wide range of biological and engineering applications such as molecular sensors for disease diagnosis, therapeutic agents for the treatment of diseases, and a vehicle for delivering therapeutics and imaging agents for theranostic applications in cells and living animals.

Abilash Gangula · Brandon Kim · Benjamin Casey ·
Allison Hamill · Hariharan Regunath ·
Anandhi Upendran

Point-of-Care Testing of COVID-19

Current Status, Clinical Impact, and Future Therapeutic Perspectives

 Springer

Abilash Gangula
Department of Radiology
School of Medicine
University of Missouri
Columbia, MO, USA

Brandon Kim
Department of Radiology
School of Medicine
University of Missouri
Columbia, MO, USA

Benjamin Casey
Department of Biological, Biomedical,
and Chemical Engineering
University of Missouri
Columbia, MO, USA

Allison Hamill
Department of Biological, Biomedical,
and Chemical Engineering
University of Missouri
Columbia, MO, USA

Hariharan Regunath
Divisions of Pulmonary and Critical Care
Medicine, and Infectious Diseases
Department of Medicine
School of Medicine
University of Missouri
Columbia, MO, USA

Anandhi Upendran
Department of Biological, Biomedical,
and Chemical Engineering
University of Missouri
Columbia, MO, USA

Office of Medical Research
School of Medicine
MU Institute for Clinical and Translation
Sciences (MU-iCATS)
University of Missouri
Columbia, MO, USA

Department of Medical Pharmacology
and Physiology
School of Medicine
University of Missouri
Columbia, MO, USA

ISSN 2191-530X   ISSN 2191-5318 (electronic)
SpringerBriefs in Applied Sciences and Technology
ISSN 2197-6740   ISSN 2197-6759 (electronic)
Nanotheranostics
ISBN 978-981-19-4956-2   ISBN 978-981-19-4957-9 (eBook)
https://doi.org/10.1007/978-981-19-4957-9

This Springer imprint is published by the registered company Springer Nature Singapore Pte Ltd.
The registered company address is: 152 Beach Road, #21-01/04 Gateway East, Singapore 189721, Singapore

Contents

About the Authors

Abilash Gangula is a postdoctoral research fellow in the Department of Radiology, University of Missouri, Columbia. He has obtained his doctoral degree from Sri Sathya Sai Institute of Higher Learning, Prasanthi Nilayam, India, on the thesis entitled "Studies on the catalytic and sensing applications of Gold Nanoparticles". His research interests include the synthesis and application of bionanoconjugates for point-of-care diagnostic platforms and targeted drug delivery for cancer therapy. e-mail: abilashg@health.missouri.edu

Brandon Kim is a student researcher in the fields of radiology and neuroscience at the University of Missouri, Columbia. He has worked on projects such as the development of a rapid COVID-19 antigen test for local use through diagnostic gold nanoparticles and is currently working on a project on the possible link between glial activation through outside stresses or immune response and retinal-degenerative diseases. e-mail: bk8br@health.missouri.edu

Benjamin Casey is an alumnus of the University of Missouri, Columbia. He obtained a Bachelor of Science in Biological Engineering. His emphasis track throughout his undergraduate career was premedicine. e-mail: bnctk8@health.missouri.edu

Allison Hamill is an alumnus of the University of Missouri, Columbia. She has a Bachelor of Science in Biomedical Engineering, emphasizing bioinformatics. With her technical and medical background in bioengineering, she now works at Epic Systems Corporation to support healthcare software across the globe.

Hariharan Regunath received the M.B.B.S. degree from The Tamil Nadu Dr. M. G. R. Medical University, India, and the M.D. degree from Manipal University, India. He received additional residency training in internal medicine and fellowship training in infectious diseases and critical care medicine at the University of Missouri, Columbia. He is currently an Assistant Professor of clinical medicine with the Department of Medicine—Divisions of Pulmonary and Critical Care Medicine, and Infectious Diseases, University of Missouri. His current interests include COVID-19,

ARDS, nosocomial infections, infective endocarditis, clinical microbiology, CMV infections, and travel medicine. e-mail: regunathh@health.missouri.edu

Anandhi Upendran is a Lead Scientist and Director of Biomedical Innovation in the office of medical research, School of Medicine, University of Missouri (MU). She received her Master's from Madras University and her Ph.D. from the Indian Institute of Science, Bangalore. She received her certificate in Regulatory Affairs (Drugs and Devices) from Regulatory Affairs Professional Society (RAPS, USA). At MU, she leads a research program focused on *"Clinical Translation of Drug Delivery and Biomedical Devices"*. She serves, trains, and educates undergraduate, graduate students, and early career scientists in biomedical innovation through the "Life Sciences Innovation and Entrepreneurship" graduate certificate program. Her team conducts the preclinical evaluation of new nanomaterials for effective translation to humans. She has published several peer-reviewed articles in high-impact journals and issued US patents. e-mail: upendrana@health.missouri.edu

Chapter 1
Point-of-Care Testing of COVID-19: Current Status, Clinical Impact, and Future Therapeutic Perspectives

Abstract The effective management of the coronavirus disease 2019 (COVID-19) pandemic depends on the bedrock of rapid and accurate testing to enable promptness in quarantining, contact tracing, epidemiologic characterization, evaluating vaccine response, and strategic decision-making. In this context, point-of-care (POC) tests are of utmost importance as they facilitate rapid and decentralized testing without much instrumentation and technical expertise. The review describes the current status of POC COVID-19 testing in three broad categories: molecular, antigen, and antibody. The advantages, limitations, and adaption of each of the three types of POC tests are discussed while highlighting their clinical impact in real-world settings. The role of POC testing for COVID-19 screening, diagnosis, and surveillance has been highlighted, focusing on recent advances in the field. The difference between POC and at-home tests is discussed while elaborating on the necessity for the latter. A spotlight on the impact of variants on the performance of COVID-19 tests is provided. The clinical impact of POC testing in hospitals with regard to improving therapeutic options, patient flow, enhancing the infection control measures, and early recruitment of patients into clinical trials is discussed. Finally, the future perspectives that will aid the research community in the development of POC tests for COVID-19 or any infectious disease, in general, are presented. Overall, we believe this review can benefit the research community as it (i) presents a comprehensive understanding of current COVID-19 POC testing methods (ii) highlights features required to transform the current tests developed during the past year as POC diagnostics, and (iii) provides insights to address the unmet challenges in the field.

Keywords COVID-19 · SARS-CoV-2 · Point-of-care · Testing · Diagnosis · Screening · Surveillance · Variants · At-home

1.1 Introduction

Severe acute respiratory syndrome coronavirus 2 (SARS-CoV-2) is one of the strains of coronavirus causing the global coronavirus disease 2019 (COVID-19) pandemic (Hu et al. 2021). In the past 2 years, COVID-19 infected more than 517 million people and caused over 6.25 million deaths(https://coronavirus.jhu.edu/). The United

© The Author(s), under exclusive license to Springer Nature Singapore Pte Ltd. 2022   1
A. Gangula et al., *Point-of-Care Testing of COVID-19*,
Nanotheranostics, https://doi.org/10.1007/978-981-19-4957-9_1

States faced the major brunt with more than 69 million cases and above 860,000 deaths (https://coronavirus.jhu.edu/). According to CDC data, COVID-19 was the third leading cause of death in the US in 2021 after heart disease and cancer (https://www.cdc.gov/media/releases/2022/s0422-third-leading-cause.html; Cynthia Cox 2021).

Despite the extraordinary rates of vaccine development and vaccination drives across the world, the magnitude of the threat posed by COVID-19 continues to remain high. Some of the factors propagating the pandemic are as follows: (1) prevailing inadequacy of vaccines in developing and underdeveloped nations; (2) vaccine hesitancy among certain individuals and groups; (3) rapid mutations in SARS-CoV-2 virus that create strains that are more contagious and virulent such as the Delta and Omicron; (4) decrease in the effectiveness of the vaccines and related immune response over time; and (5) asymptomatic spread of the infection. Hence, rapid testing with optimal accuracy remains the first critical step in containing and eliminating the spread of COVID-19 (Vandenberg et al. 2021).

Testing for COVID-19 comprises three important modalities, i.e., diagnosis, screening, and surveillance. Diagnostic tests are for personal health, focused on symptomatic and at-risk patients to aid in clinical interventions, while screening and surveillance tests are for population health (Mina and Andersen 2021). Screening tests are performed on symptomatic and asymptomatic populations to curb disease transmission by timely isolation of infectious individuals. Surveillance testing is intended to measure past exposures and current transmission profiles in representative samples of a population (asymptomatic patients, wastewater, surfaces) to understand the prevalence of COVID-19 and guide public health policy decisions (Mina and Andersen 2021). The desirable levels of key performance attributes such as sensitivity, specificity, and frequency of a COVID-19 test can vary based on its intended role (Mina and Andersen 2021). Diagnostic tests require high sensitivity and specificity to avoid false negatives and false positives. Screening tests need high specificity, moderate sensitivity, and high testing frequency as per the type of population (Mina and Andersen 2021). Surveillance tests can serve their purpose with moderate sensitivity and specificity as they are generalized to a population (Mina and Andersen 2021).

Molecular test or nucleic acid amplification test (NAAT) is the most widely used diagnostic test for COVID-19 due to its high specificity and sensitivity (Green et al. 2020). They also serve the purpose of screening and surveillance, as per the need. The process involves the detection of SARS-CoV-2 RNA in specimens from the upper or lower respiratory tract (https://www.cdc.gov/coronavirus/2019-ncov/lab/naats.html). These molecular tests can be broadly categorized into two types: (1) reverse transcription-polymerase chain reaction (RT-PCR) that involves temperature cycling-based amplification and (2) isothermal (or constant temperature) amplification (https://www.cdc.gov/coronavirus/2019-ncov/lab/naats.html). Among these, the RT-PCR test is considered the gold standard for COVID-19 (Ulinici et al. 2021; Carrie 2022). The popularity of RT-PCR is attributed to its long-standing reputation as a diagnostic assay apart from its high sensitivity, specificity, accuracy, and resilience to variants due to the use of primers for the conserved genes of SARS-CoV-2 (Carrie

2022; Valones et al. 2009; Chloe 2022). Kim et al. have performed a meta-analysis of 19 studies on the diagnostic performance of RT-PCR for COVID-19 and observed a pooled sensitivity and specificity of 89 and 99%, respectively (Kim et al. 2020a). Although sensitive, specific, and accurate, the RT-PCR is a cumbersome multi-step technique involving expensive thermocyclers and detection kits, trained personnel, and laboratory setup (Mardian et al. 2021; Munne et al. 2021; Teymouri et al. 2021). The time to receive the final results ranges from 24 h to 7 days, including the delays for collection and transportation. (Mina and Andersen 2021; Carrie 2022; CDC 2022; Song et al. 2021a; American Molecular Dx and PanDx™ 2022). Further, RT-PCR is believed to produce false-negative results in cases of improper specimen collection, processing and preservation, and technical limitations (Mardian et al. 2021; Munne et al. 2021; Teymouri et al. 2021; Chen et al. 2020). By addressing some of these limitations, the isothermal amplification technique is gaining popularity. However, like the RT-PCR, they are also often restrained to the laboratory setup (Mina and Andersen 2021; Taleghani and Taghipour 2021). Hence, low-income countries and resource-poor settings do not have easy, timely, and widespread access to such molecular tests (Aziz et al. 2020). Thus, there is a dire need to prioritize the development of point-of-care (POC) COVID-19 tests with rapid turnaround times that are easy-to-use, widely accessible, sensitive, accurate, and cost-effective.

POC tests are diagnostic tests carried out at or near the location of patient specimen collection and can provide the outcome within minutes rather than hours (Drain et al. 2014). POC tests, unlike the conventional laboratory tests, offer the advantage of performing the diagnosis in versatile settings such as physician offices, urgent care facilities, ambulances, pharmacies, school systems, nursing homes, drive-through sites, public places such as airports, grocery stores, entertainment halls, and at home (Mor and Waisman 2000; https://www.cdc.gov/coronavirus/2019-ncov/lab/point-of-care-testing.html). The World Health Organization (WHO) coined a term ASSURED that stands as an acronym for the key attributes of a POC test: Affordable, Sensitive, Specific, User-friendly, Rapid and Robust, Equipment-free and Deliverable to end-users (Drain et al. 2014). During the demanding times of the COVID-19 pandemic, each of these attributes of a POC test plays an important role in effective disease management as well as mitigating the stress on the overburdened health care system.

Rapid POC tests facilitate immediate quarantine and contact tracing measures rather than waiting for 24 h to 7 days as is the case with standard RT-PCR laboratory tests. Considering the high reproductive numbers or R_0 (average number of secondary cases an infected person can cause) for variant of concerns (VOC) such as Delta ($R_0 = 5$) (Liu and Rocklöv 2021), and Omicron ($R_0 \geq 7$) (Ashley 2022; Ito et al. n.a), rapid detections followed by prompt isolation and contact tracing play a crucial role in controlling the spread. At the beginning of the pandemic, testing for COVID-19 started with symptomatic patients at clinics and testing centers for diagnosis (Mercer and Salit 2021). However, surging case rates have necessitated the creation of many testing locations outside of the healthcare facilities (Mercer and Salit 2021). The user-friendly and equipment-free POC tests are convenient for this purpose as well as for those who avoid testing at clinics or public places due to socioeconomic and psychological factors (Perry et al. 2021).

Antigen tests are apt for POC or at-home use as it involves the detection of viral protein segments such as the nucleocapsid (N) protein or spike (S) protein using rapid and easy-to-use lateral flow immunoassay (Mardian et al. 2021). However, they suffer from inherent sensitivity issues compared to molecular tests. Therefore, these rapid antigen tests best serve the purpose of screening, rather than diagnostics. Both molecular and antigen tests can only detect an ongoing infection but cannot determine prior infections (https://www.cdc.gov/coronavirus/2019-ncov/symptoms-testing/testing.html).

Serological tests, the third type, fill this gap. Rather than detecting viral genes or proteins, these tests detect levels of antibodies (IgM, IgG) generated as a part of immune response in blood samples (Russo et al. 2020). Because IgM and IgG reportedly peak at 3 weeks from symptom onset, these are not useful as diagnostic tests, but rather serve as an epidemiological resource to evaluate the spread of COVID-19   (https://www.cdc.gov/coronavirus/2019-ncov/symptoms-testing/testing.html; Russo et al. 2020; Kevadiya et al. 2021).

The widely used three different COVID-19 tests, i.e., molecular, antigen, and serology tests play an important role in the effective management of the disease. Among several factors that influence the effectiveness of these tests, timing of the tests is an important determinant (Fig. 1.1). Molecular tests and antigen tests could diagnose the infection during the symptomatic stage of the disease when the virus is actively replicating (Falzone et al. 2021). Importantly, molecular tests are sensitive enough to detect the disease during the asymptomatic or presymptomatic stages when the viral loads are low while the antigen tests are less effective at this stage (Benda et al. 2021). Serological tests are effective 3–4 weeks after the exposure to the virus when the host's immune system starts to produce immunoglobulins. Thus, choosing an appropriate test is the key for efficient diagnosis, treatment, and containment of COVID-19.

In this review, we summarize the recent developments in POC COVID-19 tests in all three categories (molecular, antigen, and serology). We discuss the

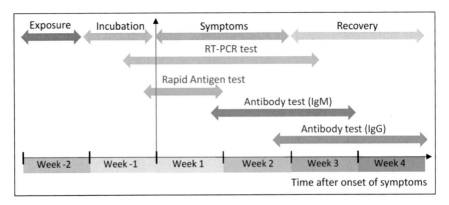

Fig. 1.1 Suitability of different types of in vitro diagnostic tests during the progression of COVID-19 infection. Excerpted from Benda et al. (2021)

diagnostic performance of several POC tests that are currently in clinical use, being developed, and commercialized across the globe, their advantages, limitations, and performance expectations to be developed as at-home tests. A spotlight on (1) at-home tests as more accessible POC tests and (2) the impact of variants on the performance of tests is provided. Finally, we have presented detailed future perspectives that support innovation, outlook, and policy decisions required for therapeutic implications, effective management of COVID-19 and future pandemics. In the past 2 years, many reviews have been published on diagnosis aspect of COVID-19 in general (Chen et al. 2020; Taleghani and Taghipour 2021; Russo et al. 2020; Kevadiya et al. 2021; Yüce et al. 2021; Pascarella et al. 2020; Kabir et al. 2021; Habli et al. 2021). A very few of them centered on the role of POC diagnosis in the management of the disease (Song et al. 2021a; Aljabali et al. 2021; Etienne et al. 2021; Krishnan et al. 2021; Valera et al. 2021). This review focuses in depth on different modalities of POC testing of COVID-19, provides an update of recent research advances in the field, presents the clinical performance of the POC tests globally, uncovers the challenges and opportunities, and evaluates the tests based on three different objectives of testing: diagnosis, screening, and surveillance.

1.2 Methods

We performed systematic and comprehensive literature research using the online databases PubMed and MEDLINE. We have reviewed all the recent articles published between January 1, 2021 and October 31, 2021. The scientific publications were retrieved from the databases using the keywords "SARS-CoV-2", "COVID-19", "Point-of-care", "Diagnosis", and "Testing". The search resulted in a total of 1087 articles after filtering for English as a language. All the articles were manually assessed based on their title and abstract to exclude duplicates and those that are outside the scope of this review. Additionally, a selection of relevant articles from November 2021 to January 2022 was also reviewed. Clinical diagnosis of COVID-19 using (1) observation of eleven common disease symptoms listed by U.S. CDC (https://www.cdc.gov/coronavirus/2019-ncov/symptoms-testing/symptoms.html) (2) Chest imaging by computed tomography (CT) and X-ray, and (3) measurement of biomarker levels in biofluids is not included in the current review as it is outside the definition of POC testing. The central theme of this review is on in vitro COVID-19 POC tests, and the readers are directed to other articles for detailed information on clinical diagnosis of COVID-19 (Mardian et al. 2021; Taleghani and Taghipour 2021; Kevadiya et al. 2021; Pascarella et al. 2020).

1.3 Choice of Clinical Specimen for COVID-19 Testing

An important prerequisite in designing a test for the detection of SARS-CoV-2 is the understanding of its structure and its concentrations in various human tissues and body fluids. SARS-CoV-2 is categorized as a Beta coronavirus. It is spherical and has a central single-stranded positive-sense RNA genome (29,881 bp in length) encapsulated by a lipid bilayer membrane (Huang et al. 2020a). Two-thirds of RNA genome consists of overlapping, ORF1ab (open reading frame) region that predominantly encodes non-structural proteins that are only produced by the virus during its replication in the host (Yadav et al. 2021a). The rest of the genome encodes the structural proteins: nucleocapsid (N) protein; membrane (M) protein; envelope (E) protein; and spike (S) protein that renders the crown-like ("corona") surface morphology. (Yadav et al. 2021a; Ke et al. 2020) The S protein mediates viral entry into the host respiratory cells by binding to the angiotensin-converting enzyme 2 (ACE2) receptors expressed on the cell membrane. The virus replicates inside the host cell and eventually disseminates via the systemic circulation to other tissue and organs. However, the viral load in each tissue is not in proportion to their degree of ACE2 expression (Trypsteen et al. 2020).

The SARS-CoV-2 virus or its components were located in several organs (pharynx, trachea, lungs, blood, heart, vessels, intestines, brain, male genitals, and kidneys), body fluids (mucus, saliva, urine, cerebrospinal fluid, semen, and breast milk), and stool of humans (Trypsteen et al. 2020; Cevik et al. 2021; Cheung et al. 2020). However, for diagnostics, swabs obtained from the nasopharynx or oropharynx are commonly used since the highest viral loads (median of 10^5 copies per ml) are observed in the upper respiratory tract in the initial phase of the infection (Mardian et al. 2021; Chen et al. 2020). But, the discomfort associated with nasopharyngeal (NP) sampling could potentially lead to inappropriate sample collection, resulting in false-negative outcomes. Saliva samples are a convenient alternative as they offer pain-free sampling and eliminate the need for trained personnel to collect samples and hence the associated exposure risk for the testing personnel (Mardian et al. 2021). Saliva-based tests have comparable or greater performance sensitivities than NP swabs (Callahan et al. 2021). Limitations include inaccuracies related to the variable viscosity of saliva posing technical issues, contamination with blood or mucus, presence of RNases and other inhibitory proteins, that could result in false negatives (Mekaliah et al. 2021; Chu et al. 2021). As a potential solution, the use of a gargled lavage or deep-throat specimen can increase the test accuracy via gargle solutions/buffers that can inactivate the RNA-degrading proteins and reduce the viscosity of the sample. Research reports suggest that the gargle lavage is as efficient and sensitive as NP swabs in the detection of COVID-19 (Zander et al. 2021; Mittal et al. 2020). For certain groups, such as infants where obtaining nasal/nasopharyngeal swabs or saliva may not be convenient or feasible, stool samples were considered as a convenient option. The diagnostic potential of stool is validated by the following research findings: i) the mean duration of SARS-CoV-2 RNA shedding was highest in the stool (up to 17.2 days) compared to the upper respiratory tract (17 days), lower respiratory tract (14.6 days), and serum (16.6 days); (Cevik et al. 2021) (2) stool showed the viral presence even before the pharyngeal specimen in one of the reported study; (Cheung et al. 2020) (3) stool PCR detected SARS-CoV-2 in cases with high clinical suspicion for COVID-19 when nasopharyngeal PCR was negative, indicating a continued but underappreciated diagnostic potential (Szymczak et al. 2020). The comparison between different clinical specimen is summarized in Table 1.1. The choice of specimen for the diagnostic test would be influenced by several factors including population type/age, availability of resources, the setting for the intended use (diagnosis or screening or prevalence/surveillance), and access to trained healthcare personnel. Hence, it is important to evaluate the performance of existing or new POC diagnostic tests in more than one patient specimen to ensure the usability and reliability of the test results under different scenarios mentioned above.

Table 1.1 Comparison of attributes of different clinical specimen for COVID-19 testing

Sample type	Collection materials	Storage temperature until testing (°C)	Recommended temperature for shipment	Viral load (Copies/mL)	References
Nasopharyngeal swabs	Dacron or polyester flocked swabs	2–8	2–8 °C if ≤5 days, −70 °C (dry ice) if >5 days	10^4–10^7	Fajnzylber et al. (2020), Wang et al. (2020)
Oropharyngeal swabs	Dacron or polyester flocked swabs	2–8	2–8 °C if ≤5 days, −70 °C (dry ice) if >5 days	10^4–10^7	Fajnzylber et al. (2020), Wang et al. (2020)
Saliva	Sterile container	2–8	2–8 °C if ≤3 days, −70 °C (dry ice) if >3 days	10^4–10^7	Williams et al. (2020), Azzi et al. (2020)
Sputum	Sterile container	2–8	2–8 °C if ≤2 days, −70 °C (dry ice) if >2 days	10^4–10^9	Fajnzylber et al. (2020)
Urine	Urine collection container	2–8	2–8 °C if ≤5 days, −70 °C (dry ice) if >5 days	10^2–10^3	Fajnzylber et al. (2020), Kim et al. (2020b)
Stool	Stool container	2–8	2–8 °C if ≤5 days, −70 °C (dry ice) if >5 days	10^6 copies/g	Kim et al. (2020b), Anjos et al. (2022)

Adapted from Nouri et al. (2021)

1.4   Mutations, Variants, and Diagnosis

The high rates of genetic mutation in SARS-CoV-2 have resulted in the emergence of many variants, the latest being Omicron and its subvariants (Lauring and Hodcroft 2021). A variant of a virus is its different strain having one or more mutations or changes in the original genetic sequence (https://www.fda.gov/medical-devices/cor onavirus-covid-19-and-medical-devices/sars-cov-2-viral-mutations-impact-covid-19-tests). For example, most variants of SARS-CoV-2 carry D614G mutation which is a point mutation or a single amino acid change (aspartic acid-to-glycine substitution) at position 614 in the spike glycoprotein of the original strain (Lauring and Hodcroft 2021). The D614G mutation was found to increase the transmissibility and infectivity of the virus (Zhang et al. 2020; Korber et al. 2020). Each of the variants that have once been a dominant strain is eventually replaced by the ensuing ones which turned powerful by natural selection. For example, in the U.S, most of the variants that were of severe threat in the past (Alpha, Beta, Gamma, Epsilon, Eta, Iota, Kappa, Zeta, Mu) have been currently classified as variants being monitored (VBM) due to their low prevalence, and only two of the variants (Delta and Omicron) are presently classified as variants of concern (VOC) due to evidence of their increased transmissibility, disease severity, and effect on treatments, testing, or vaccines   (https://www.cdc.gov/coronavirus/2019-ncov/symptoms-testing/testing.html).

The location of the mutation in the viral structure plays a decisive role in influencing results of a COVID-19 test (https://www.fda.gov/medical-devices/corona virus-covid-19-and-medical-devices/sars-cov-2-viral-mutations-impact-covid-19-tests). As per existing studies, the ORF1ab region that constitutes the bulk (~66%) of the SARS-CoV-2 genome is the most conserved, while most of the mutations in the variants are confined to the other smaller region of the virus comprising of unnamed ORFs and genes that encode the structural proteins, i.e., S, N, M, and E (Rochman et al. 2021). Thus, the performance of diagnostic or POC tests that target ORF1ab region was not significantly affected by the emergence of new variants, whereas the tests solely targeting S, N gene/protein or other genes/proteins corresponding to the non-ORF1ab region were found to underperform (West et al. 2022; Ferré et al. 2022).

1.5   POC Test Versus At-Home Test

It is important to understand the difference between POC and at-home tests as these two terms are sometimes used interchangeably. In short, all types of at-home tests come under the umbrella of POC tests, but not vice versa. Some of the important comparisons between POC and at-home tests, specifically for SARS-COV-2 detection are listed in Table 1.2. Over-the-counter (OTC) or direct-to-consumer (DTC) are at-home tests that can be purchased online or at a store without a prescription and can be self-performed. Additionally, there are prescription-based at-home test kits and at-home collection kits advised by the healthcare provider depending upon the requirement. At-home collection kits allow users to only collect samples. Later, the sample is shipped to CLIA (Clinical Laboratory Improvement Amendments of 1988)-certified laboratories for analysis and digital transfer of results to patients. Here, the user just avoids the inconvenience of going to a testing location but cannot avoid the long wait time (at least 24 h) to receive the results.

In terms of reliability of the above-mentioned tests, the trend decreases in the following order: POC tests > At-home collection kits > At-home tests. POC tests are more reliable as the sampling and test procedure is performed by trained personnel at

Table 1.2 Comparison between POC and at-home diagnostic tests

Attributes	POC diagnostic test	At-Home diagnostic test
Test type	NAAT or antigen test with turnaround time in minutes, rather than hours	Antigen test with turnaround time in minutes, rather than hours
Test location and operator	Sample collection and diagnosis are done by a trained staff/operator at or near the site of sample collection. The testing location is not restricted to clinical laboratories but includes various settings such as pharmacies, urgent care, schools, nursing home, temporary locations set by local organizations, etc. The testing site should have applied for a CLIA certificate of waiver	Sample collection and diagnosis are done by the user at home or location of his/her choice. Some tests have affiliated telehealth providers who can be approached to supervise the process
Cost	Free or covered by insurance	Most times, the insurance may not cover the cost of the at-home kit (ranging from $24–$38) purchased by the user. The cost can add up as people are advised to perform serial/multiple testing over several days to offset improper sampling techniques or increase the probability of detecting asymptomatic infection
Safety	Standard precautions are taken by trained staff during sample collection and handling	Entrusted to the discretion of the user to follow the instructions provided with the test. Some of the tests have affiliated telehealth providers to instruct the user
Reporting the results	The testing site should report the test result to the state, tribal, local, or territory health department	Some tests provide instructions for updating results with the health department. The choice is left to the user

permanent or temporary testing sites. Prescription-based at-home collection kits are relatively less reliable since inappropriate sample collection by inexpert users can produce false results. Lastly, at-home tests, though convenient and quick, the chances of false-negative or false-positive outcomes are high, due to improper sampling and processing steps. Among at-home tests, the ones based on prescription have better reliability since it is recommended by the healthcare provider.

1.6   POC Tests for COVID-19

In vitro COVID-19 tests are confirmatory tests highly specific to the structure of SARS-CoV-2 and can be primarily categorized into molecular, antigen, and antibody tests based on the target. (Fig. 1.2). Here, we discuss in detail the POC landscape of these three types of COVID-19 tests, their methodology, advantages, limitations, clinical performance, and recent developments in all three categories.

1.6.1   POC Molecular Tests for COVID-19

Reverse transcription-polymerase chain reaction (RT-PCR) and isothermal amplification are the two broad types of molecular tests for SARS-CoV-2 (https://www.cdc.gov/coronavirus/2019-ncov/lab/naats.html). Based on isothermal amplification technique, several methods have been developed such as nicking endonuclease amplification reaction (NEAR), transcription-mediated amplification (TMA), loop-mediated isothermal amplification (LAMP), helicase-dependent amplification (HDA), clustered regularly interspaced short palindromic repeats (CRISPR), and strand displacement amplification (SDA) (https://www.cdc.gov/coronavirus/2019-ncov/lab/naats.html). Among these, RT-LAMP is the most popular with an amplification rate of

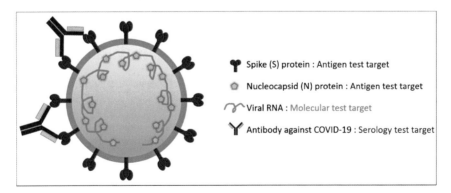

Fig. 1.2 Representation of the structure of the SARS-CoV-2 and its associated testing modalities

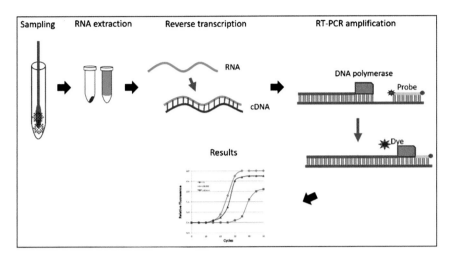

Fig. 1.3 Schematic of the workflow of RT-PCR-based SARS-CoV-2 detection

$>10^9$ copies per hour (Obande and Banga Singh 2020). CRISPR is primarily used in conjunction with the gene amplification assays to improve the sensitivity and specificity of detection. Among all the molecular tests, RT-PCR is the most widely accepted technique despite its multi-step, lengthy, expensive, and technically involved procedure (Fig. 1.3). The popularity of RT-PCR is attributed to its high sensitivity, specificity, and long-standing reputation as a reliable diagnostic method for many diseases (Valones et al. 2009).

1.6.1.1   RT-PCR-Based POC Molecular Tests

Currently, more than 250 RT-PCR-based molecular tests were approved by the U.S. FDA under emergency use authorization (EUA) for COVID-19 diagnosis (https:// www.fda.gov/medical-devices/coronavirus-disease-2019-covid-19-emergency-use-authorizations-medical-devices/in-vitro-diagnostics-euas-molecular-diagnostic-tests-sars-cov-2). Table 1.3 lists the U.S. FDA (EUA) approved RT-PCR-based COVID-19 tests that are rapid, accurate, and authorized for POC testing. Cepheid's Xpert Xpress SARS-CoV-2 test consists of a self-contained cartridge that holds all the RT-PCR reagents and fits into a GeneXpert system that performs hands-free sample-to-answer diagnosis within 30 min. The test showed excellent diagnostic accuracy in several of the studies in a clinical setting with a limit of detection of 100 copies/mL (Lee and Song 2021; Wen et al. 2021). Jian et al. performed a clinical comparison between the performance of three U.S. FDA approved POC molecular test platforms; BioFire Respiratory Panel 2.1 (RP2.1), Cobas Liat SARS-CoV-2 and Influenza A/B, and Cepheid Xpert Xpress SARS-CoV-2/Flu/RSV for their ability to detect SARS-CoV-2, especially with regard to the variant (B.1.1.7) that affected the UK in early 2021 (Jian et al. 2021). All the three tests were found to exhibit high accuracy and sensitivity in the detection of both wild type and B.1.1.7 variant with 98–100% agreement among them. However, all these POC molecular tests are authorized to be performed by trained personnel at only qualified settings holding a CLIA Certificate of Waiver, Certificate of Compliance, or Certificate of Accreditation. Similarly, Visby's rapid PCR COVID-19 test is authorized for POC testing in a healthcare setting only. The test includes several components such as a handheld test device, tube holder, buffer tube, pastette, biohazard bag and recommends the operator to have a power adapter, absorbent pads, hazardous waste disposal bin, gloves, and swabs. Thus, although RT-PCR is a popular method, the equipment, expertise, and qualified testing locations necessary to perform these tests at the POC level drive up the cost and decrease the ease of use, thereby limiting its viability as a screening test or at-home diagnostic test.

Research efforts to develop new POC-RT-PCR tests have resulted in portable designs with sensitivities and specificities greater than 95% (Paton et al. 2021; Zowawi et al. 2021). However, the total run time of the tests is long, ranging between 80 and150 min and cannot be used for low viral load samples. In an attempt to minimize the sample extraction and detection time, Razvan et al. designed Microchip-RT-PCR for SARS-CoV-2 detection wherein primers and detection probes for the target gene are preloaded in the microchip and the reaction volume was minimized to 1.2 μL resulting in a turnaround time of 30 min with similar performance as that of

Table 1.3 Selection of RT-PCR-based POC COVID-19 tests approved by the U.S. FDA (EUA)

Device	Manufacturer	Target	Specimen	LoD[a]	PPA/NPA[b]
Xpert Xpress CoV-2/Flu/RSV plus	Cepheid	SARS-CoV-2, Influenza A, B, and RSV	Upper respiratory tract	138 copies/mL	100/100%
Visby Medical COVID-19 Point-of-Care Test	Visby Medical, Inc	SARS-CoV-2	Upper respiratory tract	435 copies/swab	100/95.3%
Accula SARS-CoV-2 Test	Mesa Biotech Inc	SARS-CoV-2	Upper respiratory tract	150 copies/mL	95.8/100%
Cobas SARS-CoV-2 & Influenza A/B Nucleic Acid Test	Roche Molecular Systems, Inc	SARS-CoV-2, Influenza A, B	Upper respiratory tract	12 copies/mL	100/100%
BioFire Respiratory Panel 2.1-EZ	BioFire Diagnostics, LLC	SARS-CoV-2, other viral and bacterial respiratory organisms	Upper respiratory tract	500 copies/mL	98/100%

[a]*LoD* Limit of Detection
[b]*PPA* Positive Percent Agreement, *NPA* Negative Percent Agreement

regular RT-PCR (Cojocaru et al. 2021). However, this method has not been validated for lower respiratory specimens such as sputum that are known to harbor the viral materials for a longer duration.

Regardless of the disadvantages, RT-PCR assays may become the most reliable form of POC diagnosis with increasingly identified mutations because they use primers that target genes that are conserved in all variants (Chloe 2022; Ferré et al. 2022). For example, the in silico analysis of the commercially available RT-PCR-based ARIES® SARS-CoV-2 Assay demonstrated its effectiveness against all the current variants (Alpha, Beta, Gamma, Delta, Epsilon, Omicron) (https://www.lum inexcorp.com/aries-sars-cov-2-assay/#eua). This test targets genes in the ORF1ab as well as N genes, demonstrating high accuracy and potential to be effective against future variants as well (https://www.luminexcorp.com/aries-sars-cov-2-assay/#eua).

1.6.1.2 Isothermal Nucleic Acid Amplification-Based POC Molecular Test

One of the molecular tests that have gained in popularity since the beginning of this global pandemic is loop-mediated isothermal amplification (LAMP) or RT-LAMP. Table 1.4 lists the important differences between PCR and LAMP techniques. RT-LAMP procedure is more simplified than conventional RT-PCR as it eliminates the need for a thermocycler and uses a simple and portable heat source for gene amplification (Soroka et al. 2021). The assay uses a set of 4–6 primers and a DNA polymerase to produce strand displacement DNA at a constant temperature (60–65 °C) (Chen et al. 2020; Alkharsah 2021). The amplification readout is based on monitoring of turbidity, fluorescence, or color change. In comparison to conventional RT-PCR, RT-LAMP design is more suited for a POC application as it is rapid, easy to use, and avoids complex instruments (Chen et al. 2020).

The RT-LAMP-based POC tests that were granted EUA by the U.S. FDA are listed in Table 1.5. ID Now COVID-19 test is an instrument-based design, available by prescription only, and authorized for POC testing at CLIA testing sites only. The other two tests, Lucira and Cue, can be procured OTC and self-tested at home.

The RT-LAMP assays were found to perform well even in unprocessed samples when the test population is primarily symptomatic (Schellenberg et al. 2021). However, when dealing with blanket populations, the use of unprocessed samples has resulted in false-negative results, thereby affecting the sensitivity of the assay (Rodriguez-Mateos et al. 2021). The sample processing involves RNA extraction and purification processes ought to be performed by trained personnel thus limiting the potential of LAMP as an at-home test. Several researchers have attempted to address this shortcoming by designing an RNA extraction-free LAMP assay. Lukas et al. designed hybridization-based capture and enrichment of RNA (Bokelmann et al. 2021). The authors have used specific oligonucleotides immobilized on a magnetic bead to capture the target RNA which is then isothermally amplified followed by colorimetric detection using a smartphone application. The method is rapid, compatible with a closed reaction tube, sensitive, and specific enough to detect a single

Table 1.4 Comparison of the important attributes of LAMP and PCR techniques

Attributes	LAMP	PCR
Temperature	Isothermal reaction (60–65 °C)	Thermal cycling (multiple heating from 45 to 98 °C)
Reaction time	<1 h	~2 h
DNA extraction	Not required	Required
Primers	4–6 primers recognize 6–8 targets, extra looping primers increase sensitivity and effectiveness	2 primers recognize 2 targets
Equipment	Dry block heater/water bath	Thermocycler
Variations	Real-time LAMP, MP-LAMP, RT-LAMP	Real-time PCR, MP-PCR, RT-PCR, nested PCR, nano-PCR, long PCR, RFLP-PCR
Sensitivity	100× higher than standard PCR 100× lower than nested PCR	Depends on the variation
Read out	Naked eye, turbidimetric analysis, fluorescent detection, electrophoresis, real-time protocol	Electrophoresis, real-time protocol

Adapted from Soroka et al. (2021)

Table 1.5 Selection of RT-LAMP-based POC COVID-19 tests approved by U.S. FDA (EUA)

Device	Manufacturer	Target	Specimen	Turnaround time	LoD[a]	PPA/NPA (%)[b]
ID NOW COVID-19	Abbott Diagnostics Scarborough, Inc.	RdRp gene of SARS-CoV-2	Upper respiratory tract	13 min	125 copies/mL	100/100%
Lucira CHECK-IT COVID-19 Test Kit	Lucira Health, Inc.	RNA of N gene of SARS-CoV-2	Upper respiratory tract	30 min	2700 copies/swab	92/98%
Cue COVID-19 Test	Cue Health Inc.	The nucleic acid of SARS-CoV-2	Upper respiratory tract	20 min	1300 copies/mL	97.4/99.1%

[a]*LoD* Limit of Detection, [b]*PPA* Positive Percent Agreement, *NPA* Negative Percent Agreement

positive sample in a pool of 25 negative gargle lavage (deep-throat saliva) samples, thus facilitating high throughput to test bulk samples. David et al. exploited the binding affinity of nucleic acids to cellulose paper and designed WhoTLAMP that employs Whatman filter paper to extract RNA from saliva samples (Ng et al. 2021). The entire assay can be carried out in a 1.7 mL tube without centrifugation or any instrumentation. The Whatman paper is fixed to the bottom of the tube using an adhesive, exposed to saliva, washed a couple of times, and treated with LAMP mixture in a heat block for 20 min. The amplification of the target genome lowers the pH of the reaction medium, resulting in a color change of a pH-sensitive dye from pink to yellow. The test is sensitive enough to detect ~4 viral particles per μL of saliva within 30 min. A successful clinical validation of the test in saliva samples would make it a potential easy-to-use, sensitive, and economical at-home POC test. The test, however, uses primers that target only one gene (ORF1a) of the virus and thus can become more robust if the additional gene is included, considering the rapid mutational rates of the virus.

In a similar approach, He et al. used Si–OH-activated glass bead for RNA extraction followed by colorimetric RT-LAMP-based detection in a single tube (He et al. 2021a, 2021b). Pablo et al. designed a lab-on-a-chip platform that uses magnetic beads and an aqueous-organic immiscible interface for RNA extraction from artificial sputum before amplification and detection by RT-LAMP (Rodriguez-Mateos et al. 2021). Lee et al. have designed a paper-based multiplex POC RT-LAMP test that can differentiate and simultaneously detect SARS-CoV-2 virus, Influenza A and B viruses in saliva (Lee et al. 2021). All the steps of the assay, including RNA extraction, heating, RNA amplification, and detection, were designed to take place automatically in a sequential manner via built-in flow regulators achieving a detection limit of 50 copies/μL. The test, however, needs further optimization to minimize the run time. Deng et al. integrated RT-LAMP with a battery-powered portable cartridge that performs sample transportation, lysing, and amplification steps (Deng et al. 2021). Xun et al. developed the Scalable and Portable Testing (SPOT) RT-LAMP by designing a microfluidic chip that safely encloses all the reaction reagents and integrating it with a lateral flow assay that automates all the steps of the assay without manual involvement, thereby constituting a potential at-home POC test (Fig. 1.4) (Xun et al. 2021).

Most of the RT-LAMP assays target only one region of the viral genome resulting in low grade performance. The CDC, USA, recommends targeting at least two genes of viral RNA in a SARS-CoV-2 diagnostic test (Bhadra et al. 2021). One test that addresses this issue is the SPOT RT-LAMP, a saliva-based fluorescence assay that targets both the N and E genes of the SARS-CoV-2 separately in a small battery-powered electronic apparatus (Fig. 1.4) (Xun et al. 2021). The test reports a sensitivity of 93.3% and has the potential to be performed at home and other places without laboratory access. Song et al. have developed dual RT-LAMP (single-stage amplification, two-stage amplification) assays that use a set of custom primers targeted for ORF1ab and the N gene of SARS-CoV-2 (Song et al. 2021b). The single-stage amplification test is based on typical RT-LAMP, while the novel two-stage amplification test, termed as Penn-RAMP, has a preceding recombinase isothermal amplification step before the LAMP stage. The Penn-RAMP has 10 times more sensitivity than conventional RT-PCR. The tests can be performed at home without any technical expertise as the readout is based on color change of LCV (leuco crystal violet) dye from nearly colorless to deep violet in the presence of dsDNA. Despite qualifying to be an at-home POC diagnostic test, higher cost and longer run time make it less attractive. Additionally, the test necessitates sample processing step as its sensitivity was found to decrease in minimally processed samples, thereby warranting further optimization in the buffer composition and nature of LAMP enzymes.

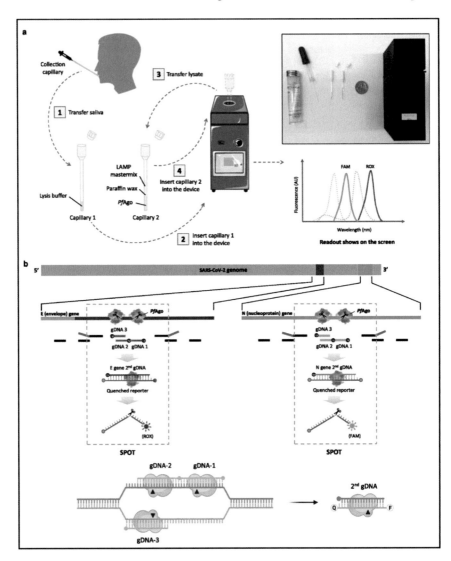

Fig. 1.4 Overall workflow using capillaries and the SPOT device. Green and red dashed lines represent excitation spectrum of FAM and ROX, respectively, while emission spectrums are shown with green and red solid lines. Inset: Device and consumables needed to run the SPOT system, including a 3D printed portable detection device, a prefabricated sample pretreatment capillary 1, a prefabricated viral RNA detection capillary 2, and collection capillaries. A quarter coin is placed for scale (diameter: 24.26 mm); b Genome map showing primers, gDNAs, and SPOT mechanism. RTLAMP primers are indicated by black rectangles. Enclosed "P" represents the phosphate group. PfAgo cleavage sites are shown in black triangles, enclosed "Q" represents quencher group and enclosed "F" represents fluorophore group. Excerpted from Xun et al. (2021)

Another limitation of the RT-LAMP method is the non-acceptable rates of false-positive diagnosis or lower specificity caused by the dependence of readout on non-specific stimuli such as turbidity, pH change, or intercalation of dyes, rather than target gene sequence. As a solution, researchers have designed LAMP assays with nucleotide probes that are specific to the target gene sequence. Choi et al. improved the specificity of RT-LAMP by integrating isothermally amplified DNA with graphene oxide-rkDNA probe that is complementary to the sequence of amplified DNA (Choi et al. 2021). In the absence of amplified target DNA, the fluorescence of rkDNA is quenched by graphene oxide while the hybridization of target DNA with the complementary rkDNA recovers the fluorescence. Ali et al. improved the specificity of RT-LAMP tests by incorporating sequence-specific QUASR (quenching of unincorporated amplification signal reporters) (Bektaş et al. 2021). Further, the sensitivity was also enhanced by using a magnetic wand for RNA enrichment. The test can also codetect influenza B along with SARS-CoV-2. Varona et al. employed molecular beacons to achieve sequence-specific detection of SARS-CoV-2 RNA (Fig. 1.5) (Varona et al. 2021). Molecular beacons (MB) are hairpin-loop-shaped oligonucleotide probes that can hybridize with the complementary target RNA sequence. The authors have designed SARS-CoV-2 RNA-specific MB with FAM on one end and biotin on the other end. The amplicon produced from the isothermal amplification process binds with the complementary sequence of MB. This complex is detected by a lateral flow strip that has anti-FAM-coated gold nanoparticles as the detection segment and streptavidin as the capture segment of the sandwich in the test zone. The method could detect the ORF1a gene of SARS-CoV-2 and found to be compatible with solid-phase microextraction to facilitate easy RNA extraction from the specimens such as human plasma, pond water, and artificial saliva. The specificity of the MB-based diagnosis is demonstrated by its ability to distinguish wild type BRAF gene of SARS-CoV-2 from its mutant (BRAF V600E), a clinically significant single-nucleotide polymorphism. Sanchita et al. have converted three RT-LAMP assays that were based on non-sequence-specific readout into a single multiplex assay that uses oligonucleotide strand displacement (OSD) probes that read in a sequence-specific manner while still retaining the colorimetric readout (Bhadra et al. 2021). The assay showed high specificity toward SARS-CoV-2 and differentiated it from other coronaviruses such as MERS-CoV and SARS-CoV Urbani. Additionally, targeting three different regions of viral RNA has resulted in high sensitivity with a detection limit of a few tens of copies of SARS-CoV-2 RNA. Ye et al. designed a recombinant FEN1-Bst DNA polymerase capable of DNA synthesis, strand displacement, and cleavage functions (Ye et al. 2021). When tested on clinical samples from 120 patients, the method exhibited 100% sensitivity and specificity in comparison with standard RT-PCR toward detection of SARS-CoV-2 (ORF1ab and N genes), rotavirus, and *Chlamydia trachomatis*. The method also facilitates viewing the endpoint with a transilluminator, thereby qualifying as a potential POC test.

Interpreting the color change in the read-out step of RT-LAMP assays is sometimes not straightforward, especially with low viral load samples. Apart from the pink (negative test) and yellow (positive test) colors observed in most of the RT-LAMP assays, the third hue of orange can be produced in samples with low viral loads

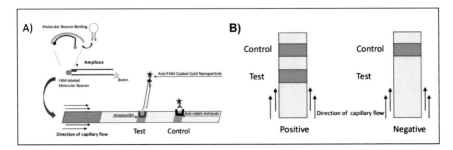

Fig. 1.5 **a** Schematic representation of MB-LAMP detection on a commercially available LFIA strip. **b** Representative illustration of LFIA strips indicating positive or negative samples. Excerpted from Varona et al. (2021). Copyright © 2021, American Chemical Society

(Aoki et al. 2021a). It is important that the interpreter needs to be aware of such possibilities and train the eyes to be vigilant. Similarly raising the temperature above 65 °C during the isothermal amplification, increasing the amplification reaction time beyond 30 min, and usage of certain primers can cause the formation of intermediate colors that are difficult to interpret (Oliveira Coelho et al. 2021). Many RT-LAMP assays do not use an internal control and may result in inaccurate results (Oliveira Coelho et al. 2021).

The real-world application of the RT-LAMP assays has been validated in numerous studies wherein the diagnostic performance of the assays showed excellent agreement with that of the gold standard RT-PCR (Wu et al. 2021; Kitajima et al. 2021; Egerer et al. 2021; García-Bernalt Diego et al. 2021; Rodriguez-Manzano et al. 2021). RT-LAMP tests can also be effective against variants as well. Saliva two-step is a colorimetric RT-LAMP that can detect the Hu-1, Alpha, Beta, and Gamma variants (Yang et al. 2021a). However, there is no conclusive data on whether this test is effective against the Delta and Omicron variants. In comparison to RT-PCR, the factors such as portability, rapid turnaround times, one-tube format, simplicity in use, and the ability to integrate with smartphones empower the RT-LAMP assays to be the most reliable POC tests for diagnosis and screening (He et al. 2021a; García-Bernalt Diego et al. 2021; Rodriguez-Manzano et al. 2021). However, sample processing requirements, sensitivity, and specificity issues limit its applications as a routine at-home test.

1.6.1.3 CRISPR-Cas-Based POC Molecular Tests

CRISPR and CRISPR-associated (Cas) protein (CRISPR/Cas)-based technology is another molecular detection method that is suited for POC diagnostic applications because of its rapidity, versatility, self-signal amplification, non-reliance on expensive instrumentation or technical expertise, high sensitivity, and specificity (Escalona-Noguero et al. 2021; Nouri et al. 2021; Kostyusheva et al. 2021; Ganbaatar and Liu 2021).

The working principle of detection of SARS-CoV-2 RNA using CRISPR-Cas technique is illustrated in Fig. 1.6. The sample collected from the patient is processed to extract the viral RNA which further undergoes nucleic acid amplification. The amplified DNA is subjected to CRISPR-Cas technique. The CRISPR-Cas system primarily consists of a Cas enzyme in complex with a guide RNA or CRISPR RNA (crRNA). The crRNA is complementary to the sequence of the target nucleic acid and leads the Cas enzyme to the target site. Cas enzyme produces site-specific DNA or RNA cleavage (cis-cleavage) followed by collateral cleavage of surrounding non-specific DNA/RNA (transcleavage) (Escalona-Noguero et al. 2021; Ganbaatar and Liu 2021). Either of these cleavage results in the release of reporter molecules that can be detected by fluorescence, colorimetric, electrochemical, or electronic readouts (Escalona-Noguero et al. 2021; Nouri et al. 2021). The most commonly used Cas enzymes for the CRISPR/Cas-based applications are Cas9, Cas12, and Cas13. The detection readout for Cas9 is based on cis-cleavage, while for Cas12 and Cas13 it is transcleavage (Escalona-Noguero et al. 2021). Mahas et al. developed a new Cas13 variant termed miniature Cas13 (mCas13) which exhibited similar sensitivity and specificity as other CRISPR-based diagnostics in the detection of SARS-CoV-2 (Mahas et al. 2021).

In COVID-19 diagnosis, the CRISPR-Cas system is primarily used in coordination with nucleic acid amplification strategies to enable sensitive and specific detection of target amplicons. For example, integration with CRISPR-Cas detection based on target-specific crRNA improves the specificity of LAMP assays. Standalone, LAMP techniques are often not specific enough to detect single-nucleotide polymorphisms, as their detection strategy is based on non-specific stimuli such as turbidity, pH change, or intercalation of dyes. Also, the LAMP technique has been combined with CRISPR-Cas to simplify the detection process of the former using naked eye, lateral flow strips, or UV lamps to enable these as potential at-home POC diagnostic tests (Ganbaatar and Liu 2021).

SHERLOCK (Specific High-Sensitivity Enzymatic Reporter UnLOCKing) and DETECTR (DNA endonuclease-targeted CRISPR transreporter) are the recently developed CRISPR-Cas-based detection systems that have been employed for highly sensitive and specific detection of SARS-CoV-2 (Broughton et al. 2020; Huang et al. 2020b; Sun et al. 2021). Both the strategies require prior isothermal recombinase polymerase amplification before the detection step. SHERLOCK employs LwaCas13a, while DETECTR uses Cas12a as specific nucleases (Ganbaatar and Liu 2021; Rahman et al. 2021) The readout for both the systems is fluorescence-based and needs a plate reader thus limiting their viability for at-home use (Ganbaatar and Liu 2021). U.S. FDA has approved EUA for two CRISPR-Cas-based COVID-19 molecular tests: Sherlock CRISPR SARS-CoV-2 Kit (Sherlock BioSciences, Inc.) and SARS-CoV-2 DETECTR Reagent Kit (Mammoth Biosciences, Inc.). However, the emergency use of these tests is limited to only authorized labs and not extended to POC settings.

Research efforts to translate CRISPR-Cas-based detection to at-home POC diagnostics has resulted in the development of SHERLOCKv2, which can simultaneously detect 4 different targets and can be integrated with gold nanoparticle (AuNP)-based

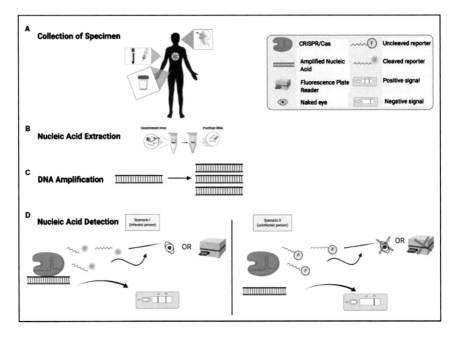

Fig. 1.6 Nucleic Acid Detection of SARS-CoV-2 Using CRISPR/Cas Assays. **a** Patient specimens can be collected from different types of clinical samples. **b** RNA is extracted from the specimen. **c** From the nucleic acid extraction, the DNA must be amplified. **d** The nucleic acid of SARS-CoV-2 can now be detected. If a person has COVID-19 (Scenario I), then the CRISPR/Cas complex will bind to the target region of the amplified nucleic acid and collateral cleavage activity can occur by cleaving the nearby fluorescence reporter nucleic acids. This can be detected by either by using the naked eye under specific light, a fluorescence plate reader, or a lateral flow assay that can indicate the presence of the virus's nucleic acid. If a person does not have COVID-19 (Scenario II), then the CRISPR/Cas complex will not bind to the target region of the amplified nucleic acid and collateral cleavage activity will not be initiated; this means that there will not be any viral signal (glow) from the sample observed by the naked eye, a plate reader, or a lateral flow assay. Excerpted from Ganbaatar and Liu (2021)

lateral flow strips for colorimetric naked-eye detection (Escalona-Noguero et al. 2021; Kostyusheva et al. 2021). The application of SHERLOCKv2 for lateral flow-based SARS-CoV-2 detection is not yet validated, but DETECTR lateral flow assay has been developed and validated in clinical samples by Broughton et al. with 95% positive percent agreement and 100% negative percent agreement in comparison to the standard RT-PCR diagnostic tests (Broughton et al. 2019). Similarly, Ackerman et al. developed a LwaCas13a SHERLOCK-based massive multiplex detection platform called CARMEN (Combinatorial Arrayed Reactions for Multiplexed Evaluation of Nucleic acid) that could detect up to 4500 nucleic acids of the different viral genome including SARS-CoV-2 (Ackerman et al. 2020). The readout is based on fluorescence microscopy and is yet to be approved as a viable POC diagnostics.

To meet the criteria of the at-home test, it is important to bypass the RNA extraction step and achieve simple and sensitive detection from the unprocessed sample in a single reaction tube (Mahas et al. 2021). Joung et al. developed SHERLOCK testing in one pot for COVID-19 (STOP covid assay) that uses LAMP to amplify viral RNA and heat-stable AapCas12b protein for CRISPR-mediated detection (Joung et al. 2020). The viral lysis, RNA extraction and concentration, RT-LAMP, and Cas12b detection are all performed in one pot at 60 °C. The assay is integrated with the lateral flow for a naked-eye readout with a detection limit up to attomolar (aM) concentration of SARS-CoV-2 nucleic acid. Similarly, Ding et al. developed an AIOD-CRISPR assay (All-In-One dual CRISPR-Cas12a) that combines all the reagents of recombinase polymerase amplification and Cas12a detection in one pot at 37 °C (Ding et al. 2020). The test uses two crRNA to increase the specificity and is based on fluorescence readout. However, it has not been integrated with lateral flow immunoassay (LFIA) to serve as an at-home diagnostics. Puig et al. developed a minimally instrumented SHERLOCK (miSHERLOCK) test that can detect most of the SARS-CoV-2 variants (Alpha, Beta, and Gamma) (de Puig et al. 2021). The design incorporates battery-powered viral RNA extraction and concentration from unprocessed saliva, one-pot SHERLOCK reactions with fluorescent output integrated with a mobile phone app for automated result interpretation to aid as at-home diagnostics (Fig. 1.7). Feng et al. integrated reverse transcription (RT), recombinase polymerase amplification (RPA), and CRISPR-Cas12a nuclease reactions into a single tube to achieve detection of SARS-CoV-2 RNA in 20 min. Importantly, the researchers demonstrated the significance of using RNase H that can decouple RNA-Cdna hybrids to free Cdna, thereby improving the kinetics and sensitivity of the RT-RPA- CRISPR-Cas12a assay (Feng et al. 2021). Azmi et al. developed CASSPIT (Cas13-Assisted Saliva-based and Smartphone Integrated Testing) that facilitates extraction-free detection of RNA from saliva samples based on LFIA readout using a smartphone application (Azmi et al. 2021). The procedure exhibited high sensitivity and has been validated in clinical samples producing 98% agreement with standard RT-PCR results.

A novel way of combining the OTC available glucometer with CRISPR detection for POC molecular testing of COVID-19 was developed by Huang et al. with a detection limit of 10 copies/μL (Huang et al. 2021a). Viral genome is placed in a solution of sucrose and is amplified through RT-RPA. The solution also contains magnetic beads coupled to invertase via an ssDNA linker. The crRNA identifies the target sequence, initiating Cas12a to cleave the target dsDNA. This further triggers transcleavage activity resulting in the breaking of ssDNA that is linking the magnetic beads and invertase. The release of invertase causes the conversion of sucrose to glucose, which can be read by the glucometer. Though the method has integrated a portable glucometer for easy readout, it has not overcome the sample processing step for extraction of viral RNA and further has not been validated in a clinical setting. Azhar et al. designed a new Cas9-based readout for COVID-19 diagnosis that utilizes FnCas9 in its FnCas9 Editor Linked Uniform Detection Assay (FELUDA) (Azhar et al. 2021). The method is integrated with a web tool called JAYTAU and is reported to show high specificity to target nucleotide sequence than commonly used Cas12 and Cas13, with a potential to detect a single mismatch in nucleotide sequence. On integration with a lateral flow readout, the assay exhibited 100% sensitivity and 97% specificity in detecting viral RNA in clinical samples within 1 h. Similarly, Marsic

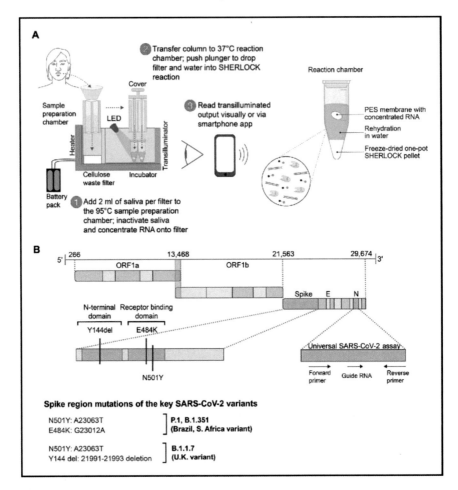

Fig. 1.7 **na** Schematic of miSHERLOCK, which integrates instrument-free viral RNA extraction and concentration from unprocessed saliva, one-pot SHERLOCK reactions that detect SARS-CoV-2 and variants, fluorescent output, and accessory mobile phone app for automated result interpretation. Step 1: The user turns on the device and introduces 4 ml of saliva into the sample preparation chamber (2 ml per filter) and adds 40 μl of 1 M DTT and 500 mM EGTA lysis buffer. Saliva flows by gravity and capillary action through a PES membrane, which accumulates and concentrates viral RNA. Step 2: The user transfers the flow columns into the reaction chamber and depresses the plunger cover to release the PES membrane and sealed stored water into freeze-dried, one pot SHERLOCK reaction pellets. Step 3: The user returns after 55 min and visualizes the assay directly or using a smartphone app that quantifies fluorescent output and automates result interpretation. The app may also be used for distributed remote result reporting. **b** SARS-CoV-2 genomic map indicating regions that are targeted in this study. The N gene target is used for a universal SARS-CoV-2 assay. SARS-CoV-2 variants are detected by targeting key mutations in the N-terminal and RBD regions of the SARS-CoV-2 spike protein, including N501Y, Y144del, and E484K. Excerpted from de Puig et al. (2021)

et al. combined Cas9-based CRISPR technology with that of LFIA and exploited the benefits of each of these two techniques to achieve molecular detection of SARS-CoV-2 with high sensitivity that is unusual for lateral flow techniques (Marsic et al. 2021). However, all these are pending clinical validation.

Although the integration of CRISPR-Cas-based diagnosis with easy read-out methods (such as glucometer or LFIA) has resulted in highly sensitive and specific detection of COVID-19 using handheld devices, the prerequisite steps of RNA extraction and amplification reduce the overall portability, turnaround time, and ease of use. Certainly, the development of the CRISPR-Cas detection test that circumvents the need for nucleic acid amplification would increase the speed and convenience of diagnosis. Fozouni et al. designed a rapid (30 min) extraction and amplification-free CRISPR-Cas13a-based test that can detect up to 100 copies/μL of SARS-CoV-2 genome using a mobile phone microscope (Fozouni et al. 2021). The method uses multiple carefully designed crRNAs and real-time monitoring of fluorescence to achieve high sensitivity and specificity. Wang et al. substituted the RT-based amplification process with a dual amplification strategy, i.e., ligation-based transcription amplification and CRISPR-Cas13a amplification for sensitive detection of SARS-CoV-2 RNA up to 82 copies (Wang et al. 2021a). Importantly, the strategy also achieves dual recognition of the SARS-CoV-2 genome via sequence-specific ligation process and crRNA, thereby producing a highly specific test that can detect even a single-nucleotide mutation. The authors demonstrated its application to profile clinically significant D614G mutation in the SARS-CoV-2 variant. Similarly, Yang et al. employed nucleic acid circuits to achieve a hybridization chain reaction (HCR)-based amplification strategy that evades expensive and tedious enzyme-dependent RT-based amplification (Yang et al. 2021b). HCR-mediated amplification is coupled with CRISPR-Cas13a-based SARS-CoV-2 detection to achieve sensitivity up to aM concentrations where the readout is based on a homemade optical-fiber evanescent wave fluorescence platform. These techniques are still in the phase of method development and have the potential to be adapted as POC devices.

1.6.1.4 Other Molecular POC Tests

Sundah et al. demonstrated a DNA-Enzyme molecular switch technology termed catalytic amplification by transition-state molecular switch (CATCH) for direct and rapid detection of SARS-CoV-2 RNA (Sundah et al. 2021). The method eliminated complex and time-consuming amplification steps, the use of primers, or any other instrumentation. It is highly sensitive with a detection limit of 8 RNA copies per mL and is versatile enough to be carried out in 96 well format or microfluidic devices with smartphone detection. A DNA complementary to the target RNA sequence is conjugated to an enzyme. The presence of trace amounts of target RNA causes it to hybridize with the complementary DNA and displace the enzyme that catalyzes the read-out reaction to produce a fluorescence signal. The high accuracy and sensitivity of the test have been validated using clinical samples of SARS-CoV-2. However, the technique suffers from the drawback of the requirement of the RNA extraction step. Barauna et al. designed an algorithm-based ultrarapid POC test that detects the presence of SARS-CoV-2 viral particles within 2 min using attenuated total reflection-Fourier transform infrared spectroscopy (ATR-FTIR) technique and genetic algorithm-linear discriminant analysis (GA-LDA) (Barauna et al. 2021). The presence of viral particles (1582 copies/mL) in the saliva was found to cause characteristic changes in the native IR spectra that can be deciphered by the algorithm. On training the algorithm using RT-PCR-based results, the method achieved good sensitivity (95%) and specificity (89%) in clinical samples. The method was proposed as an efficient tool for rapid screening at high-traffic places such as airports or public events without any additional processing steps. Since the diagnosis is completely based on the judgment of the algorithm, elaborate training based on many RT-PCR tested samples with varying demographics is required to further validate the method. Additionally, it would help make the method scientifically robust by understanding and attributing the changes in IR signatures to the structure of the virus.

1.6.2   POC Antigen Tests for COVID-19

The antigen tests for COVID-19 are primarily based on the detection of virus-specific antigens such as N or S proteins in the upper respiratory tract specimens of the patient. The techniques employed for the antigen tests include electrochemical, optical, and other immunoassays. LFIA is the most commonly used technique for POC antigen testing of SARS-CoV-2. The LFIA-based antigen tests outperform molecular tests in terms of rapid turnaround times and convenience of use.

1.6.2.1   LFIA-Based POC Antigen Testing for COVID-19

The outbreak of COVID-19 in pandemic proportions has resulted in the development and commercialization of a plethora of LFIA-based rapid antigen tests (RATs) by several pharmaceutical and biotechnology companies across the world. LFIA is a paper-based diagnostic platform designed for rapid (within 10 min) at-home and POC detection of target analytes without a need for sophisticated instrumentation, well-equipped laboratory, or skilled personnel. LFIA is a translation of traditionally used plate-based ELISA to a chromatographic format, where the capillary forces drive the movement of the test sample across various zones to produce a colorimetric readout. Gold nanoparticles (AuNP) are the most popularly used colorimetric labels in LFIA, owing to their intense color, high stability, and ease of synthesis and biofunctionalization.

   The general working principle of AuNP-based LFIA for antigen test of SARS-CoV-2 is illustrated in Fig. 1.8. The test sample containing the N or S protein of SARS-CoV-2 is introduced at the sample pad. The capillary force drives the viral proteins to the conjugate pad where they bind with the anti-N/S protein antibody (detection Ab) conjugated to AuNP. The AuNP-detection Ab-N/S protein complex flows forward toward the detection zone where it is arrested by another anti-N/S protein antibody (capture Ab) immobilized at the test lines, thus developing an intense red color characteristic of AuNP. The AuNP-detection Ab conjugate is captured by the secondary Ab to detection Ab at the control line and generates red color confirming the credibility of the assay. The appearance of red color at both test and control lines confirms the presence of SARS-CoV-2 antigen in the test sample. There are more than 18 LFIA-based RATs that are granted EUA by U.S. FDA (https://www.fda.gov/medical-devices/coronavirus-disease-2019-covid-19-emergency-use-authorizations-medical-devices/in-vitro-diagnostics-euas-antigen-diagnostic-tests-sars-cov-2). Most of them are authorized for large-scale screening of asymptomatic patients, based on visual readout, and available OTC without prescription.

Most importantly, RATs have provided the option of at-home testing and aided in meeting the overwhelming demand for COVID-19 testing in the current times of rapid spread of infection. The US government has launched a website COVIDtests.gov to request free at-home RATS delivered to their doorsteps. Table 1.6 lists the attributes and clinical performance of RATs that are approved by the U.S. FDA (EUA) for at-home testing. All, except one, can be procured OTC without a prescription. Only two of these tests (developed by Abbott) have been affiliated with a telehealth provider who would supervise the entire process of testing.

Despite the speed and convenience that they offer, RATs underperform in terms of accuracy when compared to molecular tests. RATs have lower sensitivities and may not be able to detect the infection when viral loads are low. Serial testing over several days is recommended to re-confirm a negative test result. Numerous studies across

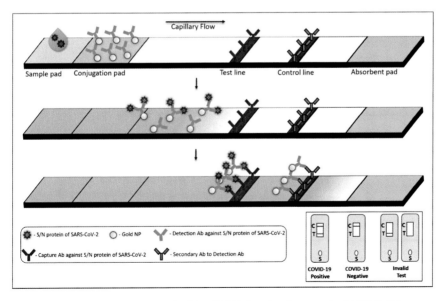

Fig. 1.8 Schematic of the working principle of LFIA-based antigen test for SARS-CoV-2

Table 1.6 Selection of At-Home COVID-19 tests approved by the U.S. FDA (EUA)

Test	Manufacturer	Technique	Attributes	Clinical performance[a]
CareStart COVID-19 Antigen Home Test	Access Bio, Inc.	Lateral flow for detection of SARS-CoV-2N protein from anterior nasal swab	Non-prescription OTC use. 10 min turnaround time, visual readout	PPA is 87%, NPA is 98%
iHealth COVID-19 Antigen Rapid Test	iHealth Labs, Inc.	Lateral flow for detection of SARS-CoV-2N protein from anterior nasal swab	Non-prescription OTC use. 15 min turnaround time, visual readout	PPA is 94.3%, NPA is 98.4%
BD Veritor At-Home COVID-19 Test	Becton, Dickinson and Company (BD)	Lateral flow for detection of SARS-CoV-2N protein from anterior nasal swab	Non-prescription OTC use. 15 min turnaround time, digital readout based on a smartphone app	PPA is 84.6%, NPA is 99.8%
SCoV-2 Ag Detect Rapid Self-Test	InBios International Inc.	Lateral flow for detection of SARS-CoV-2N protein from anterior nasal swab	Non-prescription OTC use. 20–25 min turnaround time, visual readout	PPA is 85.7%, NPA is 100%
BinaxNOW COVID-19 Antigen Self Test	Abbott Diagnostics Scarborough, Inc.	Lateral flow for detection of SARS-CoV-2N protein from anterior nasal swab	Non-prescription OTC use. 30–40 min turnaround time, visual readout	PPA is 84%, NPA is 98%
BinaxNOW COVID-19 Ag Card Home Test	Abbott Diagnostics Scarborough, Inc.	Lateral flow for detection of SARS-CoV-2N protein from anterior nasal swab	Prescription use supervised by Telehealth provider, 15 min turnaround time, visual readout	PPA is 91.7%, NPA is 100%
BinaxNOW COVID-19 Ag Card 2 Home Test	Abbott Diagnostics Scarborough, Inc.	Lateral flow for detection of SARS-CoV-2N protein from anterior nasal swab	Non-prescription OTC use supervised by Telehealth provider, 15 min turnaround time, visual readout	PPA is 84.6%, NPA is 98.5%
InteliSwab COVID-19 Rapid Test	OraSure Technologies, Inc	Lateral flow for detection of SARS-CoV-2N protein from anterior nasal swab	Non-prescription OTC use. 30 min turnaround time, visual readout	PPA is 84%, NPA is 98%

(continued)

Table 1.6 (continued)

Test	Manufacturer	Technique	Attributes	Clinical performance[a]
Celltrion DiaTrust COVID-19 Ag Home Test	Celltrion USA, Inc.	Lateral flow for detection of SARS-CoV-2N protein and RBD of S protein from the mid-turbinate swab	Non-prescription OTC use. 15 min turnaround time, visual readout	PPA is 86.7%, NPA is 99.8%
QuickVue At-Home OTC COVID-19 Test	Quidel Corporation	Lateral flow for detection of SARS-CoV-2N protein from anterior nasal swab	Non-prescription OTC use. 10 min turnaround time, visual readout	PPA is 83.5%, NPA is 99.2%
Flowflex COVID-19 Antigen Home Test	ACON Laboratories, Inc.	Lateral flow for detection of SARS-CoV-2 N protein from anterior nasal swab	Non-prescription OTC use. 15 min turnaround time, visual readout	PPA is 93%, NPA is 100%
Ellume COVID-19 Home test	Ellume Limited	Lateral flow for detection of SARS-CoV-2N protein from mid-turbinate nasal swab	Non-prescription OTC use. 15 min turnaround time, Fluorescence Analyzer	PPA is 95%, NPA is 97%

[a]*PPA* Positive Percent Agreement, *NPA* Negative Percent Agreement

the globe have compared the specificity and sensitivity of many of the commercially available RATs in comparison to the standard RT-PCR tests (Table 1.7). The cycle threshold (C_T) value of the RT-PCR is the number of PCR cycles required to amplify the viral RNA to detectable levels and thus is inversely related to the relative levels of viral RNA in a specimen (Lai and Lam 2021). The RATs which can detect the infection in samples with relatively higher C_T values are considered to possess good sensitivities.

As per the recommendations of WHO, at a minimum, RATs should exhibit a sensitivity of $\geq$ 80% and specificity of $\geq$97% in reference to RT-PCR, to be able to qualify as a reliable POC test in settings where molecular tests are not accessible or rapid turnaround times are desirable (https://www.who.int/publications/i/item/antigen-detection-in-the-diagnosis-of-sars-cov-2infection-using-rapid-immunoassays). Unfortunately, the real-world performance of most of the developed RATs (Table 1.7) falls short of WHO recommendations with regards to sensitivity, though all of them show excellent specificities (~100%). Additionally, considerable variation in sensitivity was observed (1) between a group of RATs tested on the same population; (2) for a single RAT tested on different study cohorts (Table 1.7) (Osterman et al. 2021).

Table 1.7 The diagnostic performance analysis of commercially available rapid antigen tests (RATs) for COVID-19

Test/Manufacturer	Specifications of test	Study details	Sensitivity (%)	Specificity (%)	C_T cutoff of RT-PCR	References
BinaxNOW COVID-19 Ag Card/Abbott diagnostics Scarborough, Inc	LFIA for N protein of SARS-CoV-2, authorized for symptomatic people, 15–30 min turnaround time	Nasal swab specimen of 1540 asymptomatic college students, USA	20	100	<40	Tinker et al. (2021)
		Nasal swab specimen of 2592 asymptomatic people, USA	35.8	99.8	No cutoff	Prince-Guerra, et al. (2020)
		Nasal swab specimen of 827 symptomatic people, USA	64.2	100	<34	Prince-Guerra, et al. (2020)
		Nasal swab specimen of 3302 symptomatic and asymptomatic people, USA	98.5	99.9	<35	Pilarowski et al. (2020)
		Nasal swab specimen of 2308 symptomatic and asymptomatic people, USA	81.2	> 99	≤35	Pollock et al. (2021)
LumiraDx SARS-CoV-2 Ag Test/LumiraDx UK Ltd	Microfluidic immunofluorescence assay for N protein of SARS-CoV-2, authorized for symptomatic people, 12 min, turnaround time	Nasal swab specimen of 761 symptomatic and asymptomatic people, Germany	82.2	99.3	No cutoff	Krüger et al. (2021a)
		Nasal swab specimen of 907 symptomatic and asymptomatic people, Italy	90.3	92.1	No cutoff	Bianco et al. (2021)

(continued)

Table 1.7 (continued)

Test/Manufacturer	Specifications of test	Study details	Sensitivity (%)	Specificity (%)	C_T cutoff of RT-PCR	References
Panbio™ COVID-19 Ag Rapid Test/ Panbio Ltd., Abbott Rapid Diagnostics	LFIA for N protein of SARS-CoV-2, authorized for symptomatic and asymptomatic people, 15 min turnaround time	Nasal swab specimen of 512 symptomatic and COVID positive patients, USA, UK	97.6	96.6	No cutoff	Drain et al. (2021)
		A nasopharyngeal swab of 4167 symptomatic and asymptomatic people, Italy	66.82	99.89	No cutoff	Treggiari et al. (2022)
		A nasopharyngeal swab of 433 symptomatic healthcare workers, Netherlands	86.7	100	No cutoff	Kolwijck et al. (2021)
		A nasopharyngeal swab of 332 symptomatic people, Sweden	71.8	100	<40	Nordgren et al. (2021)
		A nasal swab of 2413 symptomatic and asymptomatic people, Australia	77.3–100	99.9	No cutoff	Muhi et al. (2021)
		A nasopharyngeal swab of 3007 asymptomatic people, Canada	54.5	100	No cutoff	Shaw et al. (2021)
		A nasopharyngeal swab of 1108 symptomatic people, Germany	86.8	99.9	No cutoff	Krüger et al. (2021b)
		Combined throat/nasopharyngeal swab of 4857 symptomatic and asymptomatic people, Norway	83.8	99.9	<30	Landaas et al. (2021)

(continued)

Table 1.7 (continued)

Test/Manufacturer	Specifications of test	Study details	Sensitivity (%)	Specificity (%)	C_T cutoff of RT-PCR	References
		A nasopharyngeal swab of 535 symptomatic and asymptomatic people, Switzerland	85.5	100	No cutoff	Berger et al. (2021)
		Oropharyngeal/nasopharyngeal swab specimen of 184 symptomatic and asymptomatic people, Germany	44.6	100	No cutoff	Olearo et al. (2021)
		Nasopharyngeal swab specimen of 234 symptomatic and asymptomatic people, Italy	66	99	No cutoff	Basso et al. (2021)
		Nasopharyngeal swab specimen of 440 symptomatic children, Spain	77.7	100	No Cutoff	González-Donapetry et al. (2021)
Standard Q COVID-19 Ag Test/ SD Biosensor Inc., Roche Diagnostics	LFIA for N protein of SARS-CoV-2, authorized for symptomatic people, 15–30 min turnaround time	Nasopharyngeal swab specimen of 2375, symptomatic and asymptomatic people, Germany	68.9	99.6	No Cutoff	Holzner et al. (2021)
		Nasopharyngeal swab specimen of 529 symptomatic and asymptomatic people, Switzerland	89	99.7	No Cutoff	Jyotsna et al. (2021)
		Oropharyngeal/nasopharyngeal swab specimen of 184 symptomatic and asymptomatic people, Germany	49.4	100	No cutoff	Olearo et al. (2021)

(continued)

Table 1.7 (continued)

Test/Manufacturer	Specifications of test	Study details	Sensitivity (%)	Specificity (%)	C_T cutoff of RT-PCR	References
		Nasal swab specimen of 467 symptomatic and asymptomatic people, India	89.7	99.5	No cutoff	Jyotsna et al. 2021)
		Nasopharyngeal swab specimen of 1231 symptomatic and asymptomatic healthcare workers, Slovakia	37.8	100	No cutoff	Dankova et al. (2021)
		Nasopharyngeal swab specimen of 330 symptomatic and asymptomatic people, India	81.8	99.6	No cutoff	Gupta et al. (2020)
COVID-VIRO/ AAZ	LFIA for N protein of SARS-CoV-2, authorized for symptomatic people, 15 min turnaround time	Nasopharyngeal swab specimen of 248 symptomatic and asymptomatic people, France	97.2	100	≤37	Courtellemont et al. 2021)
		Nasal swab specimen of 234, symptomatic and asymptomatic, France	96.8	100	≤32	Cassuto et al. (2021)
CLINITEST Rapid COVID-19 Antigen Test/ Zhejiang Orient Biotech Co., Siemens	LFIA for N protein of SARS-CoV-2, authorized for symptomatic and asymptomatic people, 15 min turnaround time	Nasopharyngeal swab specimen of 270, symptomatic and asymptomatic people, Spain	75.9	100	<33	Torres et al. 2021a)
		Oropharyngeal/nasopharyngeal swab specimen of 184 symptomatic and asymptomatic people, Germany	54.9	100	No cutoff	Olearo et al. 2021)
ESPLINE SARS-CoV-2/ Fujirebio	LFIA for N protein of SARS-CoV-2, 30 min turnaround time	Nasopharyngeal swab specimen of 234 symptomatic and asymptomatic people, Italy	48	100	No cutoff	Basso et al. (2021)

(continued)

Table 1.7 (continued)

Test/Manufacturer	Specifications of test	Study details	Sensitivity (%)	Specificity (%)	C_T cutoff of RT-PCR	References
		Nasopharyngeal swab specimen of 129 symptomatic people, Japan	39.7	97	No cutoff	Aoki et al. (2021b)
		Nasopharyngeal swab specimen of 174 symptomatic people, Italy	27	>99	No cutoff	Salvagno et al. (2022)
Sienna-Clarity COVID-19 Antigen Rapid Test Cassette/ Salofa Oy	LFIA for N protein of SARS-CoV-2, authorized for symptomatic people, 10–20 min turnaround time	Nasopharyngeal swab specimen of 150 symptomatic people, France	90	100	No cutoff	Mboumba Bouassa et al. (2021)
BD Veritor System for Rapid Detection of SARS-CoV-2/Becton, Dickinson and Company	LFIA for N protein of SARS-CoV-2, authorized for symptomatic and asymptomatic people, 15 min turnaround time	Nasal swab specimen of 1384 symptomatic people, USA	66.4	98.8	No cutoff	Kilic et al. (2021)
Romed Coronavirus Ag Rapid Test Cassette/Romed	LFIA for N protein of SARS-CoV-2, 15 min turnaround time	Nasopharyngeal swab specimen of 900, symptomatic people, Netherlands	73.3	98.8	No cutoff	Koeleman et al. (2021)
MEDsan SARS-CoV-2 Antigen Rapid Test/MEDsan GmbH	LFIA for N protein of SARS-CoV-2, 15 min turnaround time	Oropharyngeal/nasopharyngeal swab specimen of 184 symptomatic and asymptomatic people, Germany	45.8	97	No Cutoff	Olearo et al. (2021)
COVID-19 Ag Respi-Strip/Coris Bioconcept	LFIA for N protein of SARS-CoV-2, 15–30 min turnaround time	Nasopharyngeal swab specimen of 484 symptomatic people, India	71.9	99.9	No cutoff	Kanaujia et al. (2021)

(continued)

Table 1.7 (continued)

Test/Manufacturer	Specifications of test	Study details	Sensitivity (%)	Specificity (%)	C_T cutoff of RT-PCR	References
		Nasopharyngeal swab specimen of 193, symptomatic and asymptomatic people, Belgium	62	100	No cutoff	Seynaeve et al. (2021)
STANDARD F COVID-19 Ag FIA/ SD Biosensor Inc.	LFIA for N protein of SARS-CoV-2, authorized for symptomatic people, 30 min turnaround time	Oropharyngeal swab specimen of 3110, symptomatic and asymptomatic people, USA	69.2	99	25–<30	Kahn et al. (2021)
Coronavirus Ag rapid test Cassette/ Zhejiang Orient Gene/Healgen Biotech	LFIA for N protein of SARS-CoV-2, authorized for symptomatic people, 15 min turnaround time	Nasopharyngeal swab specimen of 332 symptomatic people, Sweden	79.5	74.4	< 40	Nordgren et al. (2021)
BIOCREDIT COVID 19 Ag test/ Rapigen Inc	LFIA for N protein of SARS-CoV-2, 15 min turnaround time	Nasopharyngeal swab specimen of 119 symptomatic people, Germany	8.1	–	No cutoff	Kenyeres et al. (2021)
Rapid COVID-19 Antigen Test/ Healgen Scientific, LLC	LFIA for N protein of SARS-CoV-2, authorized for symptomatic people, 15 min turnaround time	Nasopharyngeal swab specimen of 193, symptomatic and asymptomatic people, Belgium	88	100	No cutoff	Seynaeve et al. (2021)

Dinnes et al. performed a meta-analysis of 78 study cohorts conducted primarily in Europe and North America to evaluate the diagnostic performance of 16 RATs and 5 molecular POC tests (Dinnes et al. 2020). The sensitivity of RATs varied significantly from 0 to 94% with an average value of 56.2%, while the specificity has less variation (90–100%) with 99.5% as the average. In contrast, rapid molecular tests showed excellent sensitivity with an average of 95.2% and comparatively less variation (68–100%), whereas the specificity varied from 95 to 99.9% with an average of 98.9%. Similar results were obtained from a meta-analysis performed by Hayer et al. evaluating clinical performance of 5 RATs (Standard Q COVID-19 Ag Test by SD Biosensor, Panbio COVID-19 Ag Test by Abbott, COVID-19 Ag Respi-Strip by Coris BioConcept, COVID-Viro by AAZ-LMB, Biocredit COVID-19 Ag by RapiGEN) in 19 studies utilizing 11,109 samples with 2509 RT-PCR-positives (Hayer et al. 2021). The sensitivity showed a discrepancy between the tests even for samples possessing high viral loads (CT value <25). The specificity varied between 92.4 and 100%. Overall, RATs by Roche Diagnostics/SD Biosensor (Standard Q) and Abbott (Panbio) had the most reliable performance with a pooled sensitivity of 82.4% and 76.9% respectively. Because of the below-par sensitivities of RATs, they are ideal for testing only a highly symptomatic population (1–7 days of infection), rather than a blanket population (https://www.who.int/publications/i/item/antigen-detection-in-the-diagnosis-of-sars-cov-2infection-using-rapid-immunoassays). While positive test results from RATs can be used to confirm ongoing infection due to excellent specificities, the negative outcome certainly cannot be used as a standalone to rule-out infections and must be confirmed by molecular tests. Despite this limitation, the high specificity of RAT can play a significant role in population screening, reduce hospitalization costs, as well as guide public health and patient management decisions by influencing prompt quarantining and contact tracing (https://www.who.int/publications/i/item/antigen-detection-in-the-diagnosis-of-sars-cov-2infection-using-rapid-immunoassays; Diel and Nienhaus 2021). Additionally, as at-home tests, RATs could diagnose infections with higher viral loads thus enabling self-monitoring and prompt isolation practices with convenience and safety as well as reducing the burden of testing on healthcare systems.

The colorimetric readout of LFIA that ultimately dictates the diagnosis and sensitivity of RATs is primarily based on two factors: (1) the efficiency of binding interaction of antigen with the antibody or antibody-nanoparticle conjugate; (2) absorption coefficient and optical properties of nanoparticle label (Razo et al. 2018). The sensitivity of RATs can be improved by optimizing these two factors with regards to type and structure of an antibody, choice and number of antigens targeted, enrichment of antigen, and choice, size, and load of the nanoparticle reporter. In this regard, Hodge et al. observed that the use of rigid antibodies, rather than flexible antibodies, would lower the incidence of undesirable overlap of the antigen-binding regions resulting in improved sensitivity in the LFIA-based detection (Hodge et al. 2021). Additionally, the use of polyclonal antibodies in LFIA is suggested to improve its performance in detecting VOCs. Polyclonal antibodies, unlike monoclonal, have binding affinity to multiple epitopes, and hence stand better equipped to detect the nucleocapsid (N) protein even if the mutation occurs at one of the immunorecognition sites (Osterman et al. 2021). The choice of target antigen may also affect the sensitivity and specificity of RAT. Moria et al. observed that tests that target spike protein are less sensitive compared to those that target nucleocapsid protein while the latter is less specific than the former (Barlev-Gross et al. 2021). Thus, the user or manufacturer of the tests must carefully choose the target antigen based on the intended use of the test. With a rapid emergence of VOCs, simultaneous detection of more than one antigen improves the accuracy and sensitivity of the assay by reducing the number of false negatives. Wang et al. designed ultrasensitive quantum dot-based LFIA for the simultaneous detection of two antigens (N and S proteins) of SARS-CoV-2 (Wang et al. 2021b). The authors have used magnetic quantum dot nanoparticle (MagTQD) as the reporter to address the lower sensitivities achieved by conventional AuNPs. The test can be performed in two modes: direct detection mode for quick and urgent screening (10 min); enrichment detection method (35 min) where in the antigens are concentrated to achieve high sensitivity in cases of samples with low viral loads. The tests exhibited high sensitivities with detection limits of 1 and 0.5 pg/mL respectively and are compatible with both saliva and nasal swab samples. Panferov et al. demonstrated that the sensitivity of LFIA can also be enhanced multi-fold by enlargement of AuNP labels postassay by deposition of silver or gold layers (Panferov et al. 2021).

1.6.2.2   Electrochemical and Optical Sensing-Based POC Antigen Testing of SARS-CoV-2

The electrochemical sensors are known to exhibit very low detection limits and can address the poor sensitivity of LFIA-based detection techniques. Significant research efforts have been directed to develop electrochemical sensors for SARS-CoV-2 detection due to their rapid response times, high sensitivity, multiplex analysis, low costs, label-free detection, and ability to be integrated into a handheld device (Hosu et al. 2018; Florea et al. 2019; Yadav et al. 2021b). The electrochemical biosensors are generally based on recognition of the target species by the bioreceptor (antigen/antibody/peptide/aptamer/DNA) immobilized on an electrode or electroactive material, where the event of target recognition results in the generation of an electronic signal proportional to the concentration of target species (Grieshaber et al. 2008; Rahmati et al. 2021; Brazaca et al. 2021).

The performance of the electrochemical sensor can be influenced by the choice of bioreceptor used in the test (Szunerits et al. 2021). Li et al. developed a multi-channel electrochemical immunosensor assay on a small disposable carbon electrode (Li et al. 2021). This is a sandwich assay involving an antispike monoclonal antibody and Horseradish peroxidase-labeled detection antibody. When tested on a pool of 79 clinical samples, the assay showed high sensitivity and specificity in comparison to ELISA with true positive and false-positive rates of 100% and 5.4%, respectively. Jiran et al. designed antibody labeled magnetic nanobeads to capture and enrich the SARS-CoV-2N protein from serum samples (Li and Lillehoj 2021). The magnetic beads also contain an HRP enzyme that can catalyze the read-out reaction resulting in an electrical signal detected by a smartphone. The complete setup was integrated on a microfluidic chip to achieve enhanced sensitivity, ease of use, minimum sample, and reagent consumption. The diagnostic performance of the method has been consistent with RT-PCR when tested on clinical samples. Torres et al. designed a handheld biosensor modified with a human ACE-2 receptor (Torres et al. 2021b). Binding of SARS-CoV-2 antigen to the biosensor is measured by an increase in the resistance to charge transfer complex of a redox probe. The biosensor exhibited sensitivity and specificity of 85.3 and 100% for nasopharyngeal/oropharyngeal swab and 100% and 86% for saliva samples, respectively.

Researchers have used Bioelectric Recognition Assay (BERA) to achieve sensitive detection of the S protein of SARS-CoV-2 (Mavrikou et al. 2021, 2020; Apostolou et al. 2021). For example, Sophie et al. have designed a cell-based biosensor method that involves the insertion of antibody for S protein into a human cell membrane by electroporation technique followed by immobilization of the cell on an electrode (Mavrikou et al. 2021). Interaction of the antibody present on the cell membrane with the S protein in the patient sample results in a measurable change in the membrane potential of the cell that is detected by a potentiometer connected to a smartphone/tablet. The biosensor exhibited a detection limit of 1 fg/mL and has a sensitivity of 92.8% and specificity of 97.8–100% compared to RT-PCR in identifying the COVID-19 positive patients. Lima et al. developed Low-cost Electrochemical Advanced Diagnostic (LEAD) that uses the strong binding affinity of human ACE2 for spike protein to develop a rapid and low-cost assay procedure (Lima et al. 2021). The test had a 100% sensitivity and specificity when tested in clinical saliva samples and 88.7% sensitivity, 86.0% specificity with nasopharyngeal/oropharyngeal samples. The result is seen through changes in the peak current of a redox probe in the solution. Similarly, Rahmati et al. designed an electrochemical sensor for S protein of SARS-CoV-2 using a screen-printed carbon electrode modified with Cu_2O nanocubes to immobilize IgG anti-SARS-CoV-2 spike antibody in an orderly fashion by using staphylococcal protein A (ProtA) (Rahmati et al. 2021). The diagnosis by the electrochemical sensor showed good agreement with that of RT-PCR when tested on both saliva and nasopharyngeal patient samples.

Despite possessing several important attributes necessary for POC devices, the electrochemical biosensors have not gained popularity in clinical diagnosis due to some of their inherent limitations (Mahshid et al. 2021). Similar to the ELISA technique, the procedure for electrochemical sensing involves several washing steps that add uncertainty and subjectivity to the assay (Kudr et al. 2021). The electrochemical methods are very sensitive to the interfering species that are generally encountered in clinical samples resulting in a low signal-to-noise ratio (Monošík et al. 2012). This necessitates additional sample processing steps that would decrease the overall convenience of this method to be developed and adapted as POC tests (Kudr et al. 2021). Additionally, these methods are often based on immobilized antibodies as biorecognition elements. Hence, like other antibody-based assays, poor characterization and quality of antibodies used, influence the reliability of the assays (Kudr et al. 2021). Chaibun et al. integrated electrochemical biosensor with isothermal amplification technique for sensitive detection of S and N genes of SARS-CoV-2 using a battery-powered potentiostat (Chaibun et al. 2021) Despite showing 100% agreement with RT-PCR results, the method has limitations such as high turnaround times (2 h) and the need for additional sample processing steps.

1.6.2.3   Other POC Antigen Tests for COVID-19

Singh et al. have reported the design of a glucometer-based aptamer competitive assay that can be adapted as a POC antigen test (Singh et al. 2021). Aptamer specific to SARS-CoV-2 protein is conjugated to a magnetic bead. The aptamer is hybridized with a complementary oligonucleotide having an invertase enzyme at its end. In the presence of the N or S protein, the aptamer prefers to hybridize with the protein thus releasing the antisense strand with the invertase enzyme. The released enzyme converts sucrose into glucose at a high turnover rate that can be easily read by a glucometer within 90 min. The test was successfully able to detect infections with 100% sensitivity and selectivity from patient saliva samples and can be potentially used for nasopharyngeal samples. However, the assay has its limitations such as finding the right aptamer, optimizing the conditions for hybridization and enzyme reaction, multiple handling steps, high cost, and the addition of detergents.

To address the limitation of below-par sensitivities of RATs, Alexandra et al. developed an optofluidic chip that can simultaneously detect and differentiate SARA-CoV-2 and influenza A antigens with a single protein sensitivity (Stambaugh et al. 2021). The design consists of the magnetic bead with a capture Ab that binds to the target antigen followed by the addition of detection Ab with a bright fluorescent reporter. The entire complex captured on the magnetic bead is then illuminated with UV light that cleaves the fluorescent reporter which is automatically transferred and detected by Antiresonant Reflecting Optical Waveguides (ARROWs). Thus, every molecule of reporter detected by ARROW corresponds to one antigen molecule captured on the magnetic bead. Currently, this is a laboratory-developed protocol and has to be optimized to translate as a commercial portable product to realize the potential of the POC application. Cennamo et al. have demonstrated a proof-of-concept for surface plasmon resonance (SPR)-based detection of S1 subunit of S protein of SARS-CoV-2 in aqueous solution (Cennamo et al. 2021). The gold layer was integrated with a molecularly imprinted polymer (MIP) that acts as a synthetic receptor, instead of an antibody, for binding to the S1 subunit of the virus. The binding event changes the refractive index at the gold-dielectric interface and is optically detected using SPR phenomena. Synthetic receptors such as MIP are less expensive and more robust toward pH and temperature changes in comparison to antibodies. To adapt these as POC diagnostics, testing on clinical samples is warranted.

1.6.3 POC Serology/Antibody Tests for COVID-19

The serological testing for SARS-CoV-2 examines the individual's whole blood/serum/plasma for the presence of IgM and IgG antibodies, that are produced as a result of immune response to the viral infection. Although it serves as an indirect and alternate method to spot COVID-19 infection, the serology tests are primarily meant for surveillance purpose. The important benefits of serology tests are (1) to determine the rate of spread of COVID-19 in a community/population and its burden; (2) to measure the effectiveness of containment, broad testing, and other policies; (3) to strategically identify and deploy those healthcare workers who developed antibodies; (4) to identify the extent of herd immunity; (5) to evaluate the vaccine-based immune response, its robustness with time, and timing of booster dose, particularly in elderly and immunocompromised individuals (Winter and Hegde 2020; Vengesai et al. 2021). The most common techniques used for serology testing of SARS-CoV-2 are enzyme-linked immunosorbent assay (ELISA), chemiluminescence immunoassay (CIA), immunofluorescence assay (IFA), and LFIA (Gong et al. 2021; Ejazi et al. 2021). Since the outbreak of COVID-19 pandemic, several serology tests based on the above techniques have been developed and granted EUA by U.S. FDA (https://www.fda.gov/medical-devices/coronavirus-disease-2019-covid-19-emergency-use-authorizations-medical-devices/eua-author ized-serology-test-performance). While the three techniques, i.e., ELISA, CIA, and IFA exhibit high sensitivity and specificity, they are not best suited for POC application due to their inherent operational procedure limitations. Compared to the other three techniques, LFIA is the most widely used for POC serology testing owing to its ease of use, rapid turnaround times, portability, low cost, and colorimetric readout without the need for instrumentation or technical expertise.

1.6.3.1 LFIA-Based POC Serology Tests for COVID-19

In terms of ease of use, LFIA best serves the purpose of POC serology testing for SARS-CoV-2. The serum levels of IgM antibody falloff quickly (average of 49 days from symptom onset) while IgG levels are detectable even after 90 days (Iyer et al. 2020). (Iyer et al. 2020). Thus, the serology tests targeting only IgM antibodies tend to have low sensitivities or high false-negative rates. As shown in Fig. 1.9, the common format in all the LFIA-based serology tests allows for total antibody detection consisting of separate test lines for IgM and IgG. The test is considered antibody positive if either of the two antibodies is detected in the whole blood/plasma/serum of the subject. The performance attributes of some of the commercially available LFIA-based serology tests are summarized in Table 1.8.

LFIA-based serology tests also suffer from poor sensitivities and low levels of agreement between one another (Lisboa Bastos et al. 2020; Yamamoto et al. 2021; Tollånes et al. 2021; Zonneveld et al. 2021). Vengesai et al. performed a systematic meta-analysis of 99 independent studies evaluating the diagnostic performance of

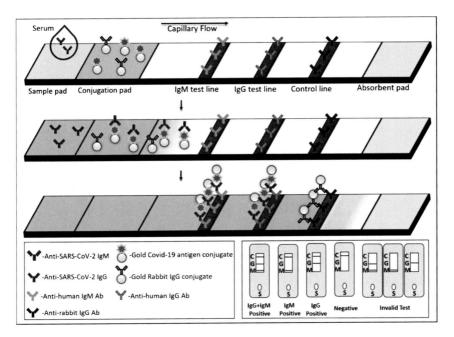

Fig. 1.9 Schematic of the working principle of LFIA-based serology test for SARS-CoV-2

serology tests for SARS-CoV-2 (Vengesai et al. 2021). In comparison to RT-PCR standard, the pooled sensitivity for LFIA-based tests are 46.3% ($n = 17$), 58.5% ($n = 16$), 68.8% ($n = 24$) for IgM, IgG, and IgM + IgG respectively, while these numbers are greater than 82% for CIA and ELISA. Particularly, serology tests do not show good sensitivity when compared to RT-PCR when tested within 14 days of onset of symptoms, since the median seroconversion time for IgM and IgG is 12 and 14 days, respectively (Pallett et al. 2021; Shyu et al. 2020). The sensitivity increases with the progression of days. Although these serology tests cannot be implemented as a diagnostic tool, they can supplement the molecular and antigen-based POC tests in improving the overall sensitivity (Espejo et al. 2020).

Research efforts were directed toward improving the sensitivity of LFIA-based serology tests since they play a critical role in epidemiologic and strategic decision-making purposes that will help control the progression of COVID-19. Kim et al. modified the universally followed LFIA format for Ab detection (Kim et al. 2021). Instead of using gold nanoparticle-antihuman IgG, the researchers used gold nanoparticle-N protein conjugate for the detection of IgG. Antihuman IgG can bind to any human Ab not specific to SARS-CoV-2, while N protein can specifically capture the IgG produced against it. By this, the sensitivity of the assay can be potentially enhanced as the interference from other human Ab in the serum can be avoided. The assay uses a vertical-flow cellulose format, and the results were found to be in concordance with the reference standard CIA. Jan et al. designed a POC magnetic immunodetection (MID) assay to improve the sensitivity and quantitative detection of antibodies

Table 1.8 Diagnostic performance analysis of commercially available LFIA-based serology tests for COVID-19

Test/Manufacturer	Specifications of test	Study details	Sensitivity (%)	Specificity (%)	Reference standard	References
SARS-CoV-2 ANTIBODY TEST/Guangzhou Wondfo Biotech Co., Ltd	LFIA, detects total antibodies (IgG + IgM) to SARS-CoV-2, 15 min turnaround time	Serum of 97 COVID-19 positive and negative control individuals, Brazil, IgM, and/or IgG	11.2	100	Fluorescence immunoassay on the same day of serology test	Pinto et al. (2020)
		Whole blood of 133 COVID-19 positive individuals, Brazil, IgM, and/or IgG	33–64	100	COVID-19 positive by RT-PCR at least 2 weeks before the serology test	Silveira et al. (2021)
		Whole blood of 100 healthcare workers, Brazil, IgM, and/or IgG	47.6	100	RT-PCR	Conte et al. (2021)
		Serum of 97 COVID-19 positive and negative control individuals, Indonesia, IgM, and/or IgG	63	95	RT-PCR	Nisa et al. (2021)
Healgen, COVID-19 IgG/IgM Rapid Test Cassette/ Healgen Scientific, LLC	LFIA, detects total antibodies (IgG + IgM) to SARS-CoV-2, 10 min turnaround time	Serum of 512 COVID-19 positive and negative control individuals, USA, IgM, and/or IgG	95.2	88.2	RT-PCR	Schuler et al. (2021)

(continued)

Table 1.8 (continued)

Test/Manufacturer	Specifications of test	Study details	Sensitivity (%)	Specificity (%)	Reference standard	References
COVID-19 IgG/IgM Rapid Test/ Prima Lab SA	LFIA, detects total antibodies (IgG + IgM) to SARS-CoV-2, 10 min turnaround time	Serum of 82 COVID-19 positive and negative control individuals, Saudi Arabia, IgM, and/or IgG	37.5	97.1	RT-PCR	Al Awaji et al. (2021)
Innovita 2019-nCoV Ab Test/ Imovita Biological Technology Co., Ltd	LFIA detects total antibodies (IgG + IgM) to SARS-CoV-2, 10–15 min turnaround	Serum of 82 COVID-19 positive and negative control individuals, Saudi Arabia, IgM, and/or IgG	41.0	85.7	RT-PCR	Al Awaji et al. (2021)
CareStart COVID-19 IgM/IgG/Access Bio	LFIA detects total antibodies (IgG + IgM) to SARS-CoV-2, 10–15 min turnaround	Serum of 512 COVID-19 positive and negative control individuals, USA, IgM, and/or IgG	96.2	89.2	RT-PCR	Schuler et al. (2021)
SARS-CoV-2 IgG/IgM Antibody test/ Joinstar Biomedical Technology Co	LFIA detects total antibodies (IgG + IgM) to SARS-CoV-2, 15 min turnaround	Serum of 97 COVID-19 positive and negative control individuals, Sweden, IgM, and/or IgG	87.0	98.0	RT-PCR	Strand et al. (2021)
COVID-19 IgG/IgM Rapid Test Cassette/ Noviral	LFIA detects total antibodies (IgG + IgM) to SARS-CoV-2, 15 min turnaround	Serum of 97 COVID-19 positive and negative control individuals, Sweden, IgM, and/or IgG	96.0	98.0	RT-PCR	Strand et al. (2021)

(continued)

Table 1.8 (continued)

Test/Manufacturer	Specifications of test	Study details	Sensitivity (%)	Specificity (%)	Reference standard	References
ZetaGene COVID-19 Rapid IgM/IgG Test/ ZetaGene Ltd	LFIA detects total antibodies (IgG + IgM) to SARS-CoV-2, 15 min turnaround	Serum of 97 COVID-19 positive and negative control individuals, Sweden, IgM, and/or IgG	85.0	98.0	RT-PCR	Strand et al. (2021)

(Pietschmann et al. 2021). An immunofiltration column was coated with recombinant SARS-CoV-2 S protein followed by incubating with serum to enrich the antibodies in it. Subsequently, a biotinylated antihuman Ab is added to the column, followed by addition of streptavidin functionalized superparamagnetic nanoparticles. Then, the column is inserted into a portable frequency mixing magnetic detection device that can quantify the magnetic particles and indirectly the antibodies in the serum. The MID assay could quantify the levels of antibody in sera of hospitalized patients (n=170) with a sensitivity of 97% and a specificity of 92%. Hung et al. coupled LFIA to a spectro-chip-based portable device that can capture the high-resolution reflectance spectrum data to quantify the serum antibody levels (Hung et al. 2021). Ibrahim et al. have simplified the design as a cost-effective LFIA assay that can be adapted as a POC testing in remote areas (Ibrahim et al. 2021). The device exhibited reproducible and accurate results when evaluated against commercial ELISA assay.

1.6.3.2 Other POC Serology Tests for COVID-19

Several alternative techniques were developed for serological testing of SARS-CoV-2 to address the limitations of the LFIA-based method. A hemagglutination test (HAT) is carried out in a well-plate containing red blood cells and the sample of interest. The assay is developed based on the quantification of the binding of the Ab with the receptor-binding domain (RBD) of the S protein of the virus that is fused with RBCs. The RBD of the spike protein of the SARS-CoV-2 virus is fused with a single-domain antibody (nanobody) IH4 to form IH4-RBD. IH4 easily binds to receptors on red blood cells (RBC), thus allowing the conjugation of RBD to RBC. When exposed to Ab specific to RBD, RBC agglutinate or clump together, due to interaction of RBD on the surface of RBC with the Ab. The agglutination or clumping of RBCs results in their homogenous dispersion while non-agglutinated RBCs tend to precipitate at the bottom of the well. Hence, a precipitate at the bottom of the well indicates the absence of Ab, while the reverse is true for Ab positive samples. The agglutination is thus visually scored to quantify the amount of Ab. A HAT qualifies as a POC test as it is rapid, low cost, and based on colorimetric readout. Alain et al. reported a HAT with a sensitivity of 90% and specificity of 99% for detection of antibodies (Townsend et al. 2021), while Kruse et al. reported sensitivity of 87% and specificity of 97% (Kruse et al. 2021a; Kruse et al. 2021b). One of the limitations of HAT include subjectivity in the visual readout as the interpretation of agglutination varies from person to person. To address this, Haecker et al. integrated the hemagglutination test with AI-based image interpretation that can automate the test result with less bias (Haecker et al. 2021).

Susanna et al. developed a bioluminescent sensor as a potential POC serology testing technique. NanoLuc (NLuc) luciferase enzyme was split into two fragments, and each of them was conjugated to a protein (S or N) of SARS-CoV-2 (Elledge et al. 2021). In the presence of Ab, the corresponding proteins hybridize with each of the two Fab arms of Ab, thus bringing the two fragments of the NLuc close enough to fuse and form a functional enzyme that produces luminescence in the presence of substrate. This assay has similar performance characteristics (sensitivity of 89% for anti-S protein antibodies and 98% for anti-N protein antibodies and a specificity of over 99% for both) to that of standard ELISA, with an overall turnaround time of just 30 min. Xu et al. developed an all-fiber Fresnel reflection microfluidic biosensor (FRMB) for rapid, easy-to-use, and sensitive detection of SARS-CoV-2 IgG and IgM antibodies with a detection limit of 0.45 ng/mL and 0.82 ng/mL respectively (Xu et al. 2021). The label-free method allows quantitative detection due to the direct relationship between Fresnel reflection light intensity and the concentration of SARS-CoV-2 IgG or IgM antibody.

Additionally, DNA nanopore sensing, a highly sensitive single-molecule detection technique has been developed for serology testing. When a potential difference is applied across a nanopore of a membrane separating a solution, translocation of charged DNA molecules present in the solution across the nanopores causes ionic current blockade that can be detected. Zhang et al. employed DNA nanopore sensing for quantitative detection of IgG and IgM antibodies in serum samples. The target antibodies in the test specimen were initially captured using a typical immunosandwich format on a magnetic bead. The detection antibody of the sandwich is conjugated to a gold nanoparticle labeled with DNA probes. The DNA probes of the sandwich complex are thermally de-hybridized from gold nanoparticles and are quantified by DNA nanopore sensing. The method could quantify the antibodies with better accuracy and dynamic range compared to ELISA and LFIA techniques and has the potential to be a POC test upon integration with microfluidic devices (Zhang et al. 2021a).

In summary, LFIA-based serology tests are easy-to-use, rapid, and economical, but suffer from poor sensitivity. To overcome this, several other methods are being developed with improved sensitivity for serology testing. However, these methods need to be further optimized to result in a user-friendly device followed by validation in large population cohorts to gain commercialization and approval by federal agencies.

1.7   Impact of Variants on the Performance of Tests

Although the prevalence of COVID-19 variants is continuously increasing across the world, their current impact on the performance of existing tests is minimal (https://www.fda.gov/medical-devices/coronavirus-covid-19-and-medical-devices/sars-cov-2-viral-mutations-impact-covid-19-tests; West et al. 2022; Rodgers et al. 2022; Omicron et al. 2022). Most NAATs are based on primers that target regions that are highly conserved between variants (such as the ORF1 region), thereby remaining less affected by the ongoing mutations. For example, the majority of RT-PCR tests were found capable of detecting the Omicron and its subvariants, the latest VOC, except for a few tests that targeted only genes of S protein and exhibited S-gene target failure (SGTF) due to S protein deletion at position 69–70 in Omicron variant (Ferré et al. 2022). To date, the U.S FDA has withdrawn approval for three molecular tests, i.e., Meridian Bioscience, Inc. Revogene SARS-CoV-2, and Applied DNA Science Linea COVID-19 Assay Kit as these tests target a portion of either N-gene or S-gene that has suffered deletion or mutation in the Omicron variant (https://www.fda.gov/medical-devices/coronavirus-covid-19-and-medical-devices/sars-cov-2-viral-mutations-impact-covid-19-tests). Similarly, diagnostic performance of RATs should remain unaffected as 90.1% are based on detecting the N protein, which has not seen mutations as much as the S protein (West et al. 2022; Ferré et al. 2022). For example, Michelena et al. have evaluated the diagnostic performance of Panbio™ COVID-19 Ag Rapid Test Device (Panbio™ RAT) with respect to standard RT-PCR in detecting Omicron variant in a prospective study involving 244 subjects (Michelena et al. 2022). Panbio™ RAT targets the N protein of SARS-CoV-2. There is a good concordance between the two types of tests as indicated by Cohen's kappa coefficient (κ) of 0.78. As expected, the median viral loads are relatively lower, or the C_T values are relatively higher for those samples where there is discordant between RT-PCR and Panbio™ RAT (Fig. 1.10). As shown in Table 1.9, the sensitivity of Panbio™ RAT increased concomitantly with the RNA load and exhibited 95.6% in specimens with viral loads ≥ 7.5 log10 copies/mL ($C_T \leq 20$). Overall, Panbio™ RAT performed well in POC diagnosis of omicron variant with a specificity and sensitivity of 100% (95% CI, 95.9–100%) and 81.8% (95% CI, 75–87.1%) respectively. Generally, the diagnostic tests based on multiple viral target sites are more likely to detect variants of SARS-CoV-2 (https://www.fda.gov/medical-devices/coronavirus-covid-19-and-medical-devices/sars-cov-2-viral-mutations-impact-covid-19-tests). Even if one of the targeted sites is mutated, the ability of the multi-target tests to detect other unmutated regions should preserve the sensitivity of the tests (https://www.fda.gov/medical-devices/coronavirus-covid-19-and-medical-devices/sars-cov-2-viral-mutations-impact-covid-19-tests).

Alternatively, testing strategies based on biomarkers that are not directly subjected to mutations would remain unaffected by the ongoing mutational changes. Alhadrami et al. have developed a FRET-based method for the detection of SARS-CoV-2 secreted proteases as biomarkers of the disease (Alhadrami et al. 2021). The authors have designed a dipeptide specific to the SARS-CoV-2 protease. The dipeptide has fluorophore at the N-terminal end whose fluorescence is quenched by the quencher at the C- terminal end. The addition of SARS-CoV-2 protease breaks the dipeptide bond and increases the fluorescence proportionately. This assay is quick, inexpensive, and is unaffected by the genomic variations of the virus as it targets the protease biomarker of SARS-CoV-2. The assay showed a detection limit of 9 ± 3 pfu (plaque-forming units) /mL when tested using in vitro cultured viral particles. However, the diagnostic performance of the test was not evaluated in detail using clinical samples consisting of COVID-19 positive and healthy controls to determine a clear distinction between the two groups. Additionally, the fluorescence readout of the assay was found to be affected by the turbidity of the sample, the presence of interfering molecules in the sample, and photobleaching of the fluorogenic substrate. Moreover, the assay needs to be performed at 37 °C for 3 h to cause maximum cleavage of dipeptide bonds.

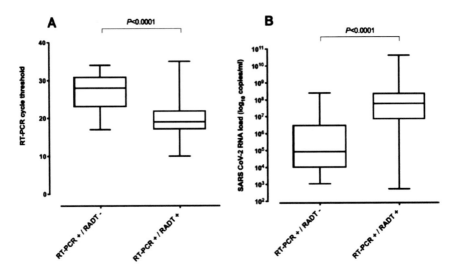

Fig. 1.10 Box-whisker plots depicting RT-PCR cycle threshold values (C_T) (**a**) and viral RNA loads (**b**) in nasopharyngeal specimens collected from COVID-19 patients infected with the Omicron variant testing either positive or negative by the Panbio™ COVID-19 Ag rapid test device (Abbott diagnostic GmbH, Jena, Germany). Excerpted from Michelena et al. (2022)

Table 1.9 Overall sensitivity of the Panbio™ COVID-19 Ag rapid test device according to the SARS-CoV-2 RNA load in nasopharyngeal specimens

RT-PCR C_T value	SARS-CoV-2 RNA load (log 10 copies/ml)	Sensitivity (95% CI)
≤20	≥7.5	95.6 (89.2–98.3)
≤25	≥5.8	92.6 (86.6–96.1)
≤30	≥4.3	87.2 (80.7–91.8)
≤35	≥2.7	81.8 (75–87.1)

Adapted from Michelena et al. (2022)

Currently, the approved POC tests in all three categories can only confirm the presence of SARS-CoV-2, but cannot identify the type of variant present in the test sample (https://www.fda.gov/medical-devices/coronavirus-covid-19-and-medical-devices/sars-cov-2-viralmutations-impact-covid-19-tests). The genomic sequencing of the sample needs to be analyzed in specialized laboratories to identify the variant. However, there are rapid tests that are currently being developed, such as SpectraLIT assay, that has demonstrated the ability to differentiate between the different variants (https://www.nsmedicaldevices.com/news/medicircle-health-spectralit-covid-19-india/). This test shines light through a swab sample and uses AI-based technology that detects patterns in the spectral information to relay test results in a matter of seconds. The design consists of small testing apparatus integrated with software, making the test portable and easy to use. Rapid tests that can differentiate between different variants will be crucial in the future for getting a quick picture of the spread of each variant in this global pandemic.

1.8   Clinical or Therapeutical Impact of POC Testing

The long turnaround times associated with centralized laboratory RT-PCR tests are considered as a major hurdle for the effective management of patients visiting the hospitals (Brendish et al. 2020). The delay in availability of the test results leads to poor patient flow with suspected patients made to wait in the assessment areas (Brendish et al. 2020). The prolonged wait times may lead to increased transmission. Such nosocomial transmission was found to be one of the major driver of infections diagnosed with the COVID-19 in the hospitals, during the first wave of the pandemic (Brendish et al. 2020). Additionally, patients awaiting the results could not participate in clinical trials aimed at improving the efficiency of the management of the pandemic. In this context, implementation of rapid and accurate POC tests in admission units and emergency departments of the hospitals would improve patient flow, reduce the rate of nosocomial transmission, facilitate timely therapeutic interventions, and maximize therapeutic benefits (Brendish et al. 2020).

Brendish et al. performed a prospective, interventional, non-randomized, controlled study to evaluate the clinical impact of molecular POC testing in comparison to that of centralized laboratory PCR test in patients with suspected COVID-19 visiting the emergency department or other acute areas of a hospital in the UK (Brendish et al. 2020). The test group comprises 499 patients that were tested with molecular POC test called the QIAstat-Dx Respiratory SARS-CoV-2 Panel. The control group includes 555 contemporaneously identified patients that were tested with laboratory PCR test. The diagnostic accuracy of both these tests was found to be similar. The primary outcome of the study was the time-to-results (time from the COVID-19 test being requested to the result being available to the clinical team) in each cohort and some of the important secondary outcomes include (1) time from admission to arrival in a definitive clinical area (i.e., a COVID-19-positive or COVID-19-negative ward) based on test results, (2) total number of bed moves prior to arrival in the correct definitive clinical area, (3) duration of hospitalization, (4) proportion of patients treated with antibiotics, (5) proportion of patients admitted to an intensive care unit (ICU), (6) in-hospital and 30-day mortality, (7) proportion of patients in each cohort enrolled into other clinical trials, and the time from admission to enrollment (Table 1.10).

The median time-to-results was 1.7 h in the POC testing group, while it was 21.3 h in the control group (Fig. 11a). The significant amount of time (19.6 h) that is saved in the POC group had a positive clinical impact as reflected in the superior outcome of other measurements. For example, the median time from admission to arrival in a definite clinical area is only 8 h in the POC testing group as opposed to 28.8 h in the control group (Fig. 11b). Similarly, the mean total number of beds moved between admission to arrival in a definitive ward was 0.9 (SD 0.5) in the POC testing

Table 1.10 Primary and secondary outcome measures

Study outcomes	POC testing	Control	Difference between groups (95% CI)[b]	P value
Time-to-results (hours)	1.7 (1.6 to 1.9)	21.3 (16.0–27.9)	19.6 (–19.0 to –20.3)	<0.0001
COVID-19 positive	197/499 (39%)	155/555 (28%)	11.5% (5.8–17.2)	0.0001
Transferred from assessment area to correct definitive clinical area on the basis of test result‡	313/428 (73%)	242/421 (57%)	15.7% (9.1–22.0)	<0.0001
Time from admission to arrival in a definitive clinical area‡, hour	8.0 (6.0–15.0)	28.8 (23.5–38.9)	−20.8 (−18.4 to −21.2)	<0.0001
Bed moves between admission and arrival in definitive clinical area[c]				
0	43/313 (14%)	0/236		
1	244/313 (78%)	163/236 (67%)		
2	26/313 (8%)	56/236 (23%)		
3	0/313	12/236 (5%)	..	.
4	0/313	4/236 (2%)		
5	0/313	1/236 (<1%)		
Mean (SD)	0.9 (0.5)	1.4 (0.7)	− 0.5 (−0.4 to− 0.6)	<0.0001
COVID-19-positive patients enrolled into other COVID-19 trials	124/197 (63%)	104/155 (67%)	− 4.2% (−14.0 to 5.9)	0.42
Time from admission to enrollment into other COVID-19 trials, days	1.0 (1.0–3.0)	3.0 (2.0–4.5)	− 2.0 (−1.0 to −2.0)	<0.0001
Antibiotics used	418/496 (84%)	387/555 (70%)	14.6% (9.5–19.5)	<0.0001
In-hospital mortality	67/494 (14%)	69/555 (12%)	1.1% (−2.9 to 5.2)	0.58
30-day mortality	80/440 (18%)	86/555 (15%)	2.6% (−2.0 to 7.3)	0.26

[a]Data are n/N (%) or median (IQR), unless otherwise specified
[b]Point-of-care testing group minus control group
[c]Assessed in patients admitted for >24 h; definitive clinical area refers to a designated COVID-19-positive or COVID-19-negative ward
Adapted from Brendish et al. (2020)

group and 1.4 (SD 0.7) in the control group (Table 1.10). Importantly, 43 (14%) of 313 patients in the POC testing group were transferred directly from the emergency department to a definitive ward bypassing the assessment area as opposed to 0 of 241in the control group. The superior patient flow and infection control measures noticed in the POC testing group would prevent the nosocomial infection among the

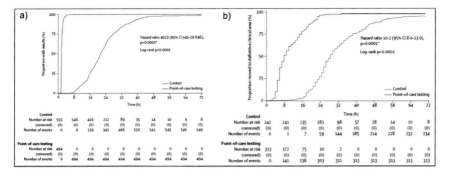

Fig. 1.11 Time-to-event curve for **a** time to results, **b** time to arrival in a definitive clinical area (i.e., COVID-19-positive or COVID-19 negative ward). Excerpted from Brendish et al. (2020). *Cox proportional hazards regression model controlling for age, sex, time of presentation, and severity of illness

patients and healthcare workers to a large extent and reduce the cost of the healthcare service for the patient (Brendish et al. 2020).

Enrollment of COVID-19 positive patients into related clinical trials is important to establish and validate the treatment efficacy of various antiviral therapies (Brendish et al. 2020). Due to rapid time-to-results in the POC testing group, COVID-19 positive patients were enrolled in the clinical trials two days earlier when compared to the control group. Early inclusion of patients is important since the antiviral therapies work best when administered during the initial phase of the disease (Brendish et al. 2020). Overall, the study by Brendish et al. strongly supports improvement of patient management, reduced transmission, and maximizing therapeutic benefit by implementation of POC testing in emergency departments and admission units of hospitals.

1.9 Testing and Treatment Perspectives

Prevention via vaccination is better than treatment, yet a significant proportion of population remain unvaccinated. Current treatment options for COVID-19 include antiviral medications and immunosuppressants. Early institution of antivirals suppresses the viral replication and thereby, inflammation. Intravenous remdesivir, the first antiviral to prove efficacy in an early clinical trial, remains as the standard of care for in-patient therapy of moderate to severe COVID-19. Recently, nirmatrelvir-ritonavir and molnupiravir have been approved as oral agents for outpatient treatment, but public awareness and access have lagged behind the actual needs. Monoclonal antibodies are also being developed and approved in parallel to the emergence of different strains, targeting patients who acquired COVID-19 and have high

Table 1.11 Temporal trend of biomarkers in COVID-19

Time period	Biomarkers
<7 days	• Total leucocyte count and lymphocyte count normal or slightly low • ↑ LDH (lactate dehydrogenase), ↑ AST (aspartate transaminase) ↑ ALT (alanine transaminase), ↑ CK (creatine kinase), ↑ CK-MB – may be early markers of severe disease and mortality
7–14 days	• Total leucocyte count & lymphocyte count progressively fall to reach nadir at 8–9 days • Thrombocytopenia may occur • ↑ IL-6 (interleukin-6), IL-10, IL-1RA, MCP-1 (monocyte chemoattractant protein 1)
>14 days	• Increasing total leucocyte count, lymphocyte and platelet count predict recovery while reducing counts predict mortality

Excerpted from Samprathi and Jayashree (2021)

risk factors. Final choice of treatment depends on the clinical severity, timeliness of diagnosis and treatment, including access to it and the choices of patients. To control the spread of a contagion, anyone with suspicious symptoms should quarantine and get tested within 2–3 days of symptom onset with a quick turn-around time of results (i.e., on the same day) to decide on treatment options. This requires streamlined messaging of the public with scientific facts in a simplified non-political language tailored to their level of understanding and concerted efforts to enhance access to testing and treatment. Future diagnostics should aim to serve such considerations, failing which pandemic disease control will not be sustained (Regunath 2022). In addition to RT-PCR diagnostic tests, that validate the presence or absence of infection, the treating physicians depend on laboratory biomarkers to obtain objective pathophysiological information for the disease management of patients that test positive for the virus. The biomarkers play a vital role in several areas of patient management such as (1) confirmation and assessment of disease severity, (2) decision-making for hospital and ICU admission, (3) identification of high-risk patients, (4) deciding therapeutic strategy, (5) examining the therapeutic response, (6) predicting the outcome, vii) decision-making for discharge from hospital and/or ICU (Samprathi and Jayashree 2021). The laboratory biomarkers studied for therapeutic management of COVID-19 include, but not limited to, hematological, inflammatory, coagulative, cardiac, and biochemical parameters (Tables 1.11 and 1.12). For asymptomatic or mildly infected patients, no blood work is recommended, while for patients in the moderate category and mild category with comorbidities, complete blood count (CBC), CRP (C-reactive protein), serum creatinine, and liver chemistries are advised during admission (Samprathi and Jayashree 2021). For patients in the severe category, levels of additional biomarkers such as prothrombin (PT), activated partial thromboplastin time (APTT), international normalized ratio (INR), serum ferritin, d-dimer, NT-pro-BNP, and troponin 1 are generally monitored (Samprathi and Jayashree 2021). Additionally, IL-6 (interleukin-6) and serial lactate levels are monitored for patients in the critical stage. The response to the therapy is monitored by repeating the CBC and CRP analysis 48–72 h after the admission (Samprathi and Jayashree 2021).

Table 1.12 Role of biomarkers in various areas of patient management

Specific role	Biomarkers
Diagnosis	Leukopenia
	Lymphopenia
	High NLR ratio ↑ LDH ↑ AST
	↑ LDH (lactate dehydrogenase)
	↑ AST (aspartate transaminase)
Assessment of severity	Lymphopenia
	Lymphocyte subsets—↓CD4+, CD8+, B, NK cells
	↑ plasma cells
	↑ NLR (neutrophil to lymphocyte ratio) and ↓ LCR (lymphocyte to CRP ratio)
	↑ IL-2R/Lymphocytes ratio
	↑ IL-6 (interleukin-6)
	↑ CRP (C-reactive protein), PCT (procalcitonin)
	↑ Ferritin
	↑ LDH
	↑ D-dimers
	↑ Specific cardiac biomarkers—CK-MB (creatine kinase), CTnT (cardiac troponin-T)
	Mb, NT-proBNP (N terminal pro brain natriuretic peptide)
Response to therapy	↓ CRP
	↓ CRP ↓ IL-6, IL-10, TNF-alpha (tumor necrosis factor-alpha), IL-2R
Prognosis	IL-6
	Ferritin
	LDH
	CRP, PCT
	Lymphocyte count
	NLR
	LCR
	Platelet count
	Specific cardiac biomarkers—CK-MB, CTnT
	Mb, NT-proBNP

Excerpted from Samprathi and Jayashree (2021)

Thus, a wide range of biomarkers are regularly analyzed to assist the physician in bed-side decision-making. In times, when the healthcare system is overwhelmed by the rapid upsurge of infections, tests that can provide the levels of biomarker with quick turnaround times would facilitate quick decision-making for the physicians and thereby significantly impact patient care. Hence, it is equally important to design and develop rapid, bed-side tests for laboratory biomarkers to enable timely treatment and monitoring of patients affected by COVID-19. Lab-on-a-chip tests that evaluate a panel of biomarkers, rather than a single biomarker, are essential.

1.10   Conclusions and Future Directions

The POC testing of COVID-19 is the effective solution to the overwhelming demand for testing at times of surge. In this regard, several COVID-19 POC tests were developed and granted EUA by U.S. FDA in three broad categories of testing: molecular, antigen, and serology. Each of the three types has their unique advantages as well as limitations. The gold standard RT-PCR-based tests are approved for POC diagnostics only in CLIA certified testing sites. The more simplified isothermal genetic amplification techniques (RT-LAMP) qualify as one of the ideal methods for POC and at-home testing. However, RT-LAMP assays often need sample processing steps and suffer from low specificity. CRISPR-Cas-based detection provides high sensitivity and specificity but requires DNA amplification steps. On the other hand, RATs and LFIA-based serology tests enable rapid at-home testing but suffer from poor sensitivities. During the past two years, a significant research effort has been directed toward addressing the limitations of all three test types resulting in innovative and improved techniques. While the current POC COVID-19 tests serve well the purpose of screening and surveillance, there is still ground to be covered with regards to the development of POC and at-home diagnostic tests that are sensitive, easy-to-use, and provide results within 10 min. Future research should aim at designing an all-in-one POC diagnostic test that could accommodate all the three test designs allowing the simultaneous detection of viral RNA, protein, and the host antibodies, while exhibiting "ASSURED" attributes recommended by WHO.

Further advancements in COVID-19 POC diagnostics can be realized by exploring new territories such as the development of tests based on exosomes, miRNA, and other unconventional molecular markers. Exosomes are extracellular nano-sized vesicles that mediate cell-to-cell communication inside the body using nucleic acid, proteins, and other molecular they carry within. Exosome-based diagnostics is gaining increasing research attention due to the following advantages they offer: i) accessible in several body fluids such as urine, plasma, saliva, and breast milk, thereby enabling easy and pain-free sampling; ii) high stability; iii) dynamic information on disease progression (Soung et al. 2017). Exosomes derived from SARS-CoV-2-infected patients were found to be carrying several molecules that are indicative of inflammation, coagulation, and immunomodulation (Barberis et al. 2021). These can potentially serve as biomarkers for the diagnosis of COVID-19 while simultaneously predicting the severity of the disease (Barberis et al. 2021). Currently, there are no COVID-19 test procedures based on exosomes. In this context, it is of importance to design POC COVID-19 diagnostic, screening, or surveillance tests based on detecting the cargo present in the exosomes derived from the patients. On a similar note, microRNA or miRNA have been recognized to play an active role in the pathogenesis of COVID-19. miRNA of the virus was found to regulate genes of the host while simultaneously suppressing the immune response (Zhang et al. 2021b). On the other hand, the plasma of COVID-19 recovered individuals was found to contain antivirus miRNA with the potential therapeutic property (Fani et al. 2021). Thus, it is worth attempting design of miRNA-based POC tests for monitoring COVID-19 and its progression. Giovannini et al. have laid out the technical road map toward achieving COVDI-19 detection just by breath analysis (Giovannini et al. 2021). Successful clinical validation of such innovative concepts would revolutionize the POC testing for COVID-19 as well as keep the nations well prepared to face future pandemics. One such painless and non-invasive technique was developed by InspectIR systems, LLC that analyzes the breath of patients to diagnose COVID-19 (https://www.fda.gov/news-events/press-announcements/cor onavirus-covid-19-update-fda-authorizes-first-covid-19-diagnostic-test-using-bre ath-samples) The technique is based on gas chromatography mass-spectrometry (GC-MS) and detects the presence of five volatile organic compound markers of COVID-19 within three minutes. In a study of 2409 individuals, the InspectIR COVID-19 Breathalyzer was found to achieve 91.2% sensitivity and 99.3% specificity. However, the test is authorized to be performed by trained personnel under the supervision of healthcare provider and is not a clinically acceptable diagnostic test. The positive test result is recommended to be confirmed with a molecular test. Further validation of this test in larger cohort studies across the globe and probably improving this as a viable at-home test would strengthen the applicability of this innovative technique.

With the growing concern of emerging variants, tests directed toward a single target site of the virus were found to be compromised by the mutational changes. The failure of certain tests in detecting omicron variants is a typical example. There is a dire need to develop POC tests, especially at-home tests, that are designed to detect multiple target sites of the virus. All the at-home RATs are antibody-based assays that target either the N or S viral proteins are subjected to mutations. Alternatively, designing at-home RATs based on aptamers that target conserved DNA sequences of the virus would be beneficial as their performance would not be affected by the ongoing mutations associated with S and N proteins. The current aptamer-based rapid tests for SARS-CoV-2 are only limited to spike protein detection, but not yet attempted for conserved RNA region (Svobodova et al. 2021; Aithal et al. 2022). Researchers can take insight from existing DNA detecting aptamer-based LFIA for other pathogens (Huang et al. 2021b). Similarly, the electrochemical biosensors known for their high sensitivities are majorly restricted to the detection of S or N proteins of SARS-CoV-2. In this context, future research efforts should be directed toward designing LFIA and electrochemical assays for the detection of genomic material of SARS-CoV-2.

Most of the existing POC tests are being validated in a single sample specimen such as a nasopharyngeal swab. Often, obtaining a nasopharyngeal swab is stressful for the patients, especially for certain populations such as infants or elders. Hence, it is important to evaluate the performance of existing or new POC diagnostic tests in more than one patient specimen (such as saliva, stool) to ensure the usability of the test under different scenarios. Another challenge faced by the patients and physicians alike is to distinguish between COVID-19, flu, RSV (respiratory syncytial virus)-based diseases, and seasonal allergies that present similar symptoms (Carrie 2022; Feng et al. 2021). Currently, there are limited approved molecular tests that are readily adaptable as POC or at-home diagnostics. There is considerable opportunity for researchers and pharmaceutical companies to develop sensitive, rapid, and cost-effective antigen tests that can reliably differentiate between these respiratory viruses.

Additional feature that will benefit the community is authentic reporting of infections. Currently, there are no streamlined ways for a direct update of results of at-home tests to local health departments by the users. All at-home tests should have a mandatory affiliation with telehealth providers who can ensure to report the test details to the public health departments. An integrated feature for direct reporting to telehealth providers or public health departments is essential for all at-home tests. This would help to realize the true infection numbers that reflect the percent positive rates to help nations and policymakers better understand the rate of spread of COVID-19.

POC tests are ideal for entry screening and public health screening as they enable prompt isolation of infectious individuals. Similarly, for surveillance purpose, POC tests facilitate easy and rapid analysis of representative samples of a population. However, the health departments and travel regulatory bodies of many countries still warrant molecular tests for screening purposes due to their high sensitivity and specificity. Among molecular tests, RT-PCR is the generally recommended and accepted test. This can be attributed to the fact that all the other methods are based on recently emerging technologies and are fast-tracked under EUA without rigorous vetting while RT-PCR has a long track record of being a diagnostic method for several metabolic, oncologic, degenerative, and infectious diseases (Valones et al. 2009). COVID-19 can be better controlled if the evaluation of the tests by the regulatory bodies and the public is based on the intended use rather than analytical performance alone. Unlike diagnostic tests, screening and surveillance tests can serve their purpose even with deviations from 100% sensitivity and specificity by compensating with attributes such as speed, convenience, and accessibility (Mina and Andersen 2021). Hence, the establishment of independent performance-analysis metrics and regulatory approval mechanisms for diagnostic and screening/surveillance tests would accelerate the deployment of more rapid POC tests to the public domain, thereby aiding in better management of COVID-19 and future pandemics.

Apart from public health screening and laboratory diagnostic settings, routine use of rapid POC tests in emergency and admission units of hospitals would have a significant clinical impact in terms of mitigating nosocomial transmission and providing prompt therapeutic interventions. Thus, resources, regulations, and research innovations should be directed toward supporting the development and regular implementation of accurate and rapid POC COVID-19 tests in the hospitals. Additionally, it is of vital research importance to develop rapid and bed-side test panels that can accurately quantify the levels of laboratory biomarkers that would help the physician in efficient and timely management of patients.

Encouragingly, the COVID-19 testing landscape in the U.S.A is more accommodative with the choice of tests for international travelers. People can opt for any antigen or molecular test (not limiting to RT-PCR) that is approved by the appropriate authority of the nation they are traveling from. Importantly, at-home tests that are approved under EUA are also accepted if they have an affiliated telehealth service that can supervise testing (Valones et al. 2009). The confidence shown on rapid POC and at-home tests by one of the most developed countries in the world would encourage other nations to follow suit and adopt POC/at-home tests for largescale and rapid screening at crowded public gatherings. The twin purposes of POC/at-home tests, i.e., i) enable decentralized testing for early detection and better control of the disease, ii) reduce the economic and procedural burden of testing, would be better served by bringing POC tests to the forefront of the testing landscape for not just COVID-19, but for other infectious diseases and future pandemics.

COVID-19 testing is going to stay with us for some time at least. Going forward, it is a combination of the following listed factors that would promote the use of POC tests in various settings of public and social life: (1) tailoring the design and development of tests based on the intended role: diagnosis/screening/surveillance; (2) establishment of less stringent or quicker regulatory approval mechanism for screening/surveillance tests as compared to diagnostic tests; (3) focused and innovative research practices aimed at improving the sensitivity, accuracy, and speed of POC tests for COVID-19 diagnosis; (4) introduction of exclusive funding programs to improve the field of POC testing and bring them to the desks of primary care physicians; (5) effective collaboration between universities and industries to translate promising POC research outcomes into hands-on devices.

Acknowledgements The authors would like to acknowledge the School of Medicine for providing small pilot funding for COVID-19 test development, and Department of Biological and Biomedical Engineering capstone project funding for undergraduate students.

References

Ackerman CM, Myhrvold C, Thakku SG, Freije CA, Metsky HC, Yang DK, Ye SH, Boehm CK, Kosoko-Thoroddsen T-SF, Kehe J, Nguyen TG, Carter A, Kulesa A, Barnes JR, Dugan VG, Hung DT, Blainey PC, Sabeti PC (2020) Massively multiplexed nucleic acid detection with Cas13. Nature 582(7811):277–282

Aithal S, Mishriki S, Gupta R, Sahu RP, Botos G, Tanvir S, Hanson RW, Puri IK (2022) SARS-CoV-2 detection with aptamer-functionalized gold nanoparticles. Talanta 236:122841

Al Awaji NN, Ahmedah HT, Alsaady IM, Bawaked RA, Alraey MA, Alasiri AA, Alfaifi AM, Alshehri HA, Alserihi R, Yasin EB (2021) Validation and performance comparison of two SARS-CoV-2 IgG/IgM rapid tests. Saudi J Biol Sci 28(6):3433–3437

Alhadrami HA, Hassan AM, Chinnappan R, Al-Hadrami H, Abdulaal WH, Azhar EI, Zourob M (2021) Peptide substrate screening for the diagnosis of SARS-CoV-2 using fluorescence resonance energy transfer (FRET) assay. Mikrochim Acta 188(4):137–137

Aljabali AAA, Pal K, Serrano-Aroca A, Takayama K, Dua K, Tambuwala MM (2021) Clinical utility of novel biosensing platform: diagnosis of coronavirus SARS-CoV-2 at point of care. Mater Lett 304:130612–130612

Alkharsah KR (20121) Laboratory tests for the detection of SARS-CoV-2 infection: basic principles and examples. German Med Sci: GMS e-Journal 19:Doc06

Anjos D, Fiaccadori FS, Servian CdP, da Fonseca SG, Guilarde AO, Borges MASB, Franco FC, Ribeiro BM, Souza M (2022) SARS-CoV-2 loads in urine, sera and stool specimens in association with clinical features of COVID-19 patients. J Clin Virol Plus 2(1):100059

Aoki K, Nagasawa T, Ishii Y, Yagi S, Kashiwagi K, Miyazaki T, Tateda K (2021b) Evaluation of clinical utility of novel coronavirus antigen detection reagent, Espline® SARS-CoV-2. J Infect Chemother: off J Jpn Soc Chemother 27(2):319–322

Aoki MN, de Oliveira Coelho B, Góes LGB, Minoprio P, Durigon EL, Morello LG, Marchini FK, Riediger IN, do Carmo Debur M, Nakaya HI, Blanes L (2021a) Colorimetric RT-LAMP SARS-CoV-2 diagnostic sensitivity relies on color interpretation and viral load. Sci Rep 11(1):9026

Apostolou T, Kyritsi M, Vontas A, Loizou K, Hadjilouka A, Speletas M, Mouchtouri V, Hadjichristodoulou C (2021) Development and performance characteristics evaluation of a new Bioelectric Recognition Assay (BERA) method for rapid Sars-CoV-2 detection in clinical samples. J Virol Methods 293:114166

Ashley B (2022) Omicron is the dominant COVID variant for two reasons. https://vitals.sutterhea lth.org/omicron-is-the-us-dominant-covid-variant-for-two-reasons/#:~:text=Omicron%20is% 20highly%20catchy.%20Here%E2%80%99s%20how%20we%20know.,The%20Omicron%20v ariant%20has%20Ro%207.0%20or%20greater. Accessed 14 Feb 2022

Azhar M, Phutela R, Kumar M, Ansari AH, Rauthan R, Gulati S, Sharma N, Sinha D, Sharma S, Singh S, Acharya S, Sarkar S, Paul D, Kathpalia P, Aich M, Sehgal P, Ranjan G, Bhoyar RC, Singhal K, Lad H, Patra PK, Makharia G, Chandak GR, Pesala B, Chakraborty D, Maiti S (2021) Rapid and accurate nucleobase detection using FnCas9 and its application in COVID-19 diagnosis. Biosens Bioelectron 183:113207

Aziz AB, Raqib R, Khan WA, Rahman M, Haque R, Alam M, Zaman K, Ross AG (2020) Integrated control of COVID-19 in resource-poor countries. Int J Infect Dis 101:98–101

Azmi I, Faizan MI, Kumar R, Raj Yadav S, Chaudhary N, Kumar Singh D, Butola R, Ganotra A, Datt Joshi G, Deep Jhingan G, Iqbal J, Joshi MC, Ahmad T (2021) A Saliva-based RNA extraction-free workflow integrated with Cas13a for SARS-CoV-2 detection. Front Cellular and Infect Microbiol 11(144)

Azzi L, Carcano G, Gianfagna F, Grossi P, Gasperina DD, Genoni A, Fasano M, Sessa F, Tettamanti L, Carinci F, Maurino V, Rossi A, Tagliabue A, Baj A (2020) Saliva is a reliable tool to detect SARS-CoV-2. J Infect 81(1):e45–e50

Barauna VG, Singh MN, Barbosa LL, Marcarini WD, Vassallo PF, Mill JG, Ribeiro-Rodrigues R, Campos LCG, Warnke PH, Martin FL (2021) Ultrarapid on-site detection of SARS-CoV-2 infection using simple ATR-FTIR spectroscopy and an analysis algorithm: high sensitivity and specificity. Anal Chem 93(5):2950–2958

Barberis E, Vanella VV, Falasca M, Caneapero V, Cappellano G, Raineri D, Ghirimoldi M, De Giorgis V, Puricelli C, Vaschetto R, Sainaghi PP, Bruno S, Sica A, Dianzani U, Rolla R, Chiocchetti A, Cantaluppi V, Baldanzi G, Marengo E, Manfredi M (2021) Circulating exosomes are strongly involved in SARS-CoV-2 infection. 8

Barlev-Gross M, Weiss S, Ben-Shmuel A, Sittner A, Eden K, Mazuz N, Glinert I, Bar-David E, Puni R, Amit S, Kriger O, Schuster O, Alcalay R, Makdasi E, Epstein E, Noy-Porat T, Rosenfeld R, Achdout H, Mazor O, Israely T, Levy H, Mechaly A (2021) Spike vs nucleocapsid SARS-CoV-2 antigen detection: application in nasopharyngeal swab specimens. Anal Bioanal Chem 413(13):3501–3510

Basso D, Aita A, Padoan A, Cosma C, Navaglia F, Moz S, Contran N, Zambon CF, Maria Cattelan A, Plebani M (2021) Salivary SARS-CoV-2 antigen rapid detection: a prospective cohort study. Clinica chimica acta; Int J Clin Chem 517:54–59

Bektaş A, Covington MF, Aidelberg G, Arce A, Matute T, Núñez I.; Walsh J, Boutboul D, Delaugerre C, Lindner AB, Federici F, Jayaprakash AD (2021) Accessible LAMP-enabled rapid test (ALERT) for detecting SARS-CoV-2. Viruses 13(5)

Benda A, Zerajic L, Ankita A, Cleary E, Park Y, Pandey S (2021) COVID-19 testing and diagnostics: a review of commercialized technologies for cost, convenience and quality of tests. Sensors 21(19):6581. https://creativecommons.org/licenses/by/4.0/

Berger A, Nsoga MTN, Perez-Rodriguez FJ, Aad YA, Sattonnet-Roche P, Gayet-Ageron A, Jaksic C, Torriani G, Boehm E, Kronig I, Sacks JA, de Vos M, Bausch FJ, Chappuis F, Renzoni A, Kaiser L, Schibler M, Eckerle I (2021) Diagnostic accuracy of two commercial SARS-CoV-2 antigen-detecting rapid tests at the point of care in community-based testing centers. PLoS ONE 16(3):e0248921

Bhadra S, Riedel TE, Lakhotia S, Tran ND, Ellington AD (2021) High-surety isothermal amplification and detection of SARS-CoV-2. mSphere 6(3)

Bianco G, Boattini M, Barbui AM, Scozzari G, Riccardini F, Coggiola M, Lupia E, Cavallo R, Costa C (2021) Evaluation of an antigen-based test for hospital point-of-care diagnosis of SARS-CoV-2 infection. J Clin Virol: off Publ Pan Am Soc Clin Virol 139:104838

Bokelmann L, Nickel O, Maricic T, Pääbo S, Meyer M, Borte M, Riesenberg S (2021) Point-of-care bulk testing for SARS-CoV-2 by combining hybridization capture with improved colorimetric LAMP. Nat Commun 12(1):1467

Brazaca LC, dos Santos PL, de Oliveira PR, Rocha DP, Stefano JS, Kalinke C, Abarza Muñoz RA, Bonacin JA, Janegitz BC, Carrilho E (2021) Biosensing strategies for the electrochemical detection of viruses and viral diseases—A review. Anal Chim Acta 1159:338384

Brendish NJ, Poole S, Naidu VV, Mansbridge CT, Norton NJ, Wheeler H, Presland L, Kidd S, Cortes NJ, Borca F, Phan H, Babbage G, Visseaux B, Ewings S, Clark TW (2020) Clinical impact of molecular point-of-care testing for suspected COVID-19 in hospital (COV-19POC): a prospective, interventional, non-randomised, controlled study. Lancet Respir Med 8(12):1192–1200

Broughton JP, Deng X, Yu G, Fasching CL, Servellita V, Singh J, Miao X, Streithorst JA, Granados A, Sotomayor-Gonzalez A, Zorn K, Gopez A, Hsu E, Gu W, Miller S, Pan C-Y, Guevara H, Wadford DA, Chen JS, Chiu CY (2020) CRISPR–Cas12-based detection of SARS-CoV-2. Nat Biotechnol 38(7):870–874

Broughton JP, Deng X, Yu G, Fasching C. L, Singh J, Streithorst J, Granados A, Sotomayor-Gonzalez A, Zorn K, Gopez A, Hsu E, Gu W, Miller S, Pan C-Y, Guevara H, Wadford D, Chen J, Chiu CY (2020) Rapid Detection of 2019 Novel Coronavirus SARS-CoV-2 Using a CRISPR-based DETECTR Lateral Flow Assay. *medRxiv* **2020**, 2020.03.06.20032334.

Callahan C, Ditelberg S, Dutta S, Littlehale N, Cheng A, Kupczewski K, McVay D, Riedel S, Kirby JE, Arnaout R (2021) Saliva is Comparable to nasopharyngeal swabs for molecular detection of SARS-CoV-2. Microbiol Spectrum 9(1):e0016221

Carrie M (2022) Which COVID-19 test should you get? https://www.yalemedicine.org/news/which-covid-test-is-accurate. Accessed 14 Feb 2022

Cassuto NG, Gravier A, Colin M, Theillay A, Pires-Roteira D, Pallay S, Serreau R, Hocqueloux L, Prazuck T (2021) Evaluation of a SARS-CoV-2 antigen-detecting rapid diagnostic test as a self-test: Diagnostic performance and usability. J Med Virol 93(12):6686–6692

CDC COVID-19 testing: what you need to know? https://www.cdc.gov/coronavirus/2019-ncov/symptoms-testing/testing.html. Accessed 8 Feb 2022

CDC COVDI-19—What you need to know about variants. https://www.cdc.gov/coronavirus/2019-ncov/variants/about-variants.html. Accessed 14 Feb 2022

CDC COVID-19—Guidance for SARS-CoV-2 rapid testing performed in point-of-care settings. https://www.cdc.gov/coronavirus/2019-ncov/lab/point-of-care-testing.html. Accessed 14 Feb 2022

CDC COVID-19—Nucleic acid amplification tests (NAATs). https://www.cdc.gov/coronavirus/2019-ncov/lab/naats.html. Accessed 14 Feb 2022

CDC COVID-19—Symptoms of COVID-19. https://www.cdc.gov/coronavirus/2019-ncov/symptoms-testing/symptoms.html. Accessed 14 Feb 2022

CDC COVID-19 was third leading cause of death in U.S. https://www.cdc.gov/media/releases/2022/s0422-third-leading-cause.html. Accessed May 11th

Cennamo N, D'Agostino G, Perri C, Arcadio F, Chiaretti G, Parisio EM, Camarlinghi G, Vettori C, Di Marzo F, Cennamo R, Porto G, Zeni L (2021) Proof of concept for a quick and highly sensitive on-site detection of SARS-CoV-2 by plasmonic optical fibers and molecularly imprinted polymers. Sensors 21(5):1681

Cevik M, Tate M, Lloyd O, Maraolo AE, Schafers J, Ho A (2021) SARS-CoV-2, SARS-CoV, and MERS-CoV viral load dynamics, duration of viral shedding, and infectiousness: a systematic review and meta-analysis. Lancet Microbe 2(1):e13–e22

Chaibun T, Puenpa J, Ngamdee T, Boonapatcharoen N, Athamanolap P, O'Mullane AP, Vongpunsawad S, Poovorawan Y, Lee SY, Lertanantawong B (2021) Rapid electrochemical detection of coronavirus SARS-CoV-2. Nat Commun 12(1):802

Chen Q, He Z, Mao F, Pei H, Cao H, Liu X (2020) Diagnostic technologies for COVID-19: a review. RSC Adv 10(58):35257–35264

Cheung KS, Hung IFN, Chan PPY, Lung KC, Tso E, Liu R, Ng YY, Chu MY, Chung TWH, Tam AR, Yip CCY, Leung KH, Fung AY, Zhang RR, Lin Y, Cheng HM, Zhang AJX, To KKW, Chan KH, Yuen KY, Leung WK (2020) Gastrointestinal manifestations of SARS-CoV-2 infection and virus load in fecal samples from a hong kong cohort: systematic review and meta-analysis. Gastroenterology 159(1):81–95

Chloe K (2022) Covid-19 variants: will the reliability of tests be affected? https://www.medicaldevice-network.com/features/covid-19-variant-tests/. Accessed 14 Feb 2022

Choi MH, Lee J, Seo YJ (2021) Combined recombinase polymerase amplification/rkDNA–graphene oxide probing system for detection of SARS-CoV-2. Anal Chim Acta 1158:338390

Chu AW-H, Yip CC-Y, Chan W-M, Ng AC-K, Chan DL-S, Siu RH-P, Chung CYT, Ng JP-L, Kittur H, Mosley GL, Poon RW-S, Chiu RY-T, To KK-W (2021) Evaluation of an automated high-throughput liquid-based RNA extraction platform on pooled nasopharyngeal or saliva specimens for SARS-CoV-2 RT-PCR. Viruses 13(4):615

Cojocaru R, Yaseen I, Unrau PJ, Lowe CF, Ritchie G, Romney MG, Sin DD, Gill S, Slyadnev M (2021) Microchip RT-PCR detection of nasopharyngeal SARS-CoV-2 samples. J Mol Diagn 23(6):683–690

Conte DD, Carvalho JMA, de Souza Luna LK, Faíco-Filho KS, Perosa AH, Bellei N (2021) Comparative analysis of three point-of-care lateral flow immunoassays for detection of anti-SARS-CoV-2 antibodies: data from 100 healthcare workers in Brazil. Braz J Microbiol [publication of the Brazilian Society for Microbiology] 52(3):1161–1165

FDA. Coronavirus (COVID-19) update: FDA authorizes first COVID-19 diagnostic test using breath samples. https://www.fda.gov/news-events/press-announcements/coronavirus-covid-19-update-fda-authorizes-first-covid-19-diagnostic-test-using-breath-samples. Accessed May 11th

Courtellemont L, Guinard J, Guillaume C, Giaché S, Rzepecki V, Seve A, Gubavu C, Baud K, Le Helloco C, Cassuto GN, Pialoux G, Hocqueloux L, Prazuck T (2021) High performance of a novel antigen detection test on nasopharyngeal specimens for diagnosing SARS-CoV-2 infection. J Med Virol 93(5):3152–3157

Cynthia Cox KA (2021) COVID-19 is the number one cause of death in the U.S. in early 2021. https://www.kff.org/coronavirus-covid-19/issue-brief/covid-19-is-the-number-one-cause-of-death-in-the-u-s-in-early-2021/

Dankova Z, Novakova E, Skerenova M, Holubekova V, Lucansky V, Dvorska D, Brany D, Kolkova Z, Strnadel J, Mersakova S, Janikova K, Samec M, Pokusa M, Petras M, Sarlinova M, Kasubova I, Loderer D, Sadlonova V, Kompanikova J, Kotlebova N, Kompanikova A, Hrnciarova M, Stanclova A, Antosova M, Dzian A, Nosal V, Kocan I, Murgas D, Krkoska D, Calkovska A, Halasova E

(2021) Comparison of SARS-CoV-2 detection by rapid antigen and by three commercial RT-qPCR tests: a study from Martin University Hospital in Slovakia. Int J Environ Res Public Health 18(13)

de Lima LF, Ferreira AL, Torres MDT, de Araujo WR, de la Fuente-Nunez C (2021) Minute-scale detection of SARS-CoV-2 using a low-cost biosensor composed of pencil graphite electrodes. Proc Natl Acad Sci U S A 118(30):e2106724118

de Oliveira Coelho B, Sanchuki HBS, Zanette DL, Nardin JM, Morales HMP, Fornazari B, Aoki MN, Blanes L (2021) Essential properties and pitfalls of colorimetric reverse transcription loop-mediated isothermal amplification as a point-of-care test for SARS-CoV-2 diagnosis. Mol Med 27(1):30

de Puig HD, Lee RA, Najjar D, Tan X, Soenksen LR, Angenent-Mari NM, Donghia NM, Weckman NE, Ory A, Ng CF, Nguyen PQ, Mao AS, Ferrante TC, Lansberry G, Sallum H, Niemi J, Collins JJ (2021) Minimally instrumented SHERLOCK (miSHERLOCK) for CRISPR-based point-of-care diagnosis of SARS-CoV-2 and emerging variants. Sci Adv 7(32):eabh2944. https://creativecommons.org/licenses/by-nc/4.0/

Deng H, Jayawardena A, Chan J, Tan SM, Alan T, Kwan P (2021) An ultra-portable, self-contained point-of-care nucleic acid amplification test for diagnosis of active COVID-19 infection. Sci Rep 11(1):15176

Diel R, Nienhaus A (2021) Point-of-care COVID-19 antigen testing in German emergency rooms— A cost-benefit analysis. Pulmonology

Ding X, Yin K, Li Z, Lalla RV, Ballesteros E, Sfeir MM, Liu C (2020) Ultrasensitive and visual detection of SARS-CoV-2 using all-in-one dual CRISPR-Cas12a assay. Nat Commun 11(1):4711

Dinnes J, Deeks JJ, Adriano A, Berhane S, Davenport C, Dittrich S, Emperador D, Takwoingi Y, Cunningham J, Beese S, Dretzke J, Ferrante di Ruffano L, Harris IM, Price MJ, Taylor-Phillips S, Hooft L, Leeflang MM, Spijker R, Van den Bruel A (2020) Rapid, point-of-care antigen and molecular-based tests for diagnosis of SARS-CoV-2 infection. Cochrane Database Syst Rev 8(8):Cd013705

Drain PK, Hyle EP, Noubary F, Freedberg KA, Wilson D, Bishai WR, Rodriguez W, Bassett IV (2014) Diagnostic point-of-care tests in resource-limited settings. Lancet Infect Dis 14(3):239–249

Drain PK, Ampajwala M, Chappel C, Gvozden AB, Hoppers M, Wang M, Rosen R, Young S, Zissman E, Montano M (2021) A rapid, high-sensitivity SARS-CoV-2 nucleocapsid immunoassay to aid diagnosis of acute COVID-19 at the point of care: a clinical performance study. Infect Dis Therapy 10(2):753–761

American Molecular Dx PanDx™ (2022) COVID-19 same day PCR test. https://amdxinc.com/covid-19-same-day-pcr-test/#:~:text=The%20average%20turn-around-time%2C%20depending%20on%20outbreak%20location%2C%20can,affordability%20also%20hinders%20proper%20testing%20of%20all%20populations. Accessed 14 Feb 2022

Egerer R, Edel B, Löffler B, Henke A, Rödel J (2021) Performance of the RT-LAMP-based eazyplex® SARS-CoV-2 as a novel rapid diagnostic test. J Clin Virol: off Publ Pan Am Soc Clin Virol 138:104817

Ejazi SA, Ghosh S, Ali N (2021) Antibody detection assays for COVID-19 diagnosis: an early overview. Immunol Cell Biol 99(1):21–33

Elledge SK, Zhou XX, Byrnes JR, Martinko AJ, Lui I, Pance K, Lim SA, Glasgow JE, Glasgow AA, Turcios K, Iyer NS, Torres L, Peluso MJ, Henrich TJ, Wang TT, Tato CM, Leung KK, Greenhouse B, Wells JA (2021) Engineering luminescent biosensors for point-of-care SARS-CoV-2 antibody detection. Nat Biotechnol 39(8):928–935

Escalona-Noguero C, López-Valls M, Sot B (2021) CRISPR/Cas technology as a promising weapon to combat viral infections. BioEssays: News Rev Mol Cell Dev Biol 43(4):e2000315

Espejo AP, Akgun Y, Al Mana AF, Tjendra Y, Millan NC, Gomez-Fernandez C, Cray C (2020) Review of current advances in serologic testing for COVID-19. Am J Clin Pathol 154(3):293–304

Etienne EE, Nunna BB, Talukder N Wang Y, Lee ES (2021) COVID-19 biomarkers and advanced sensing technologies for point-of-care (POC) diagnosis. Bioengineering (Basel) 8(7)

Fajnzylber J, Regan J, Coxen K, Corry H, Wong C, Rosenthal A, Worrall D, Giguel F, Piechocka-Trocha A, Atyeo C, Fischinger S, Chan A, Flaherty KT, Hall K, Dougan M, Ryan ET, Gillespie E, Chishti R, Li Y, Jilg N, Hanidziar D, Baron RM, Baden L, Tsibris AM, Armstrong KA, Kuritzkes DR, Alter G, Walker BD, Yu X, Li JZ, Abayneh BA, Allen P, Antille D, Balazs A, Bals J, Barbash M, Bartsch Y, Boucau J, Boyce S, Braley J, Branch K, Broderick K, Carney J, Chevalier J, Choudhary MC, Chowdhury N, Cordwell T, Daley G, Davidson S, Desjardins M, Donahue L, Drew D, Einkauf K, Elizabeth S, Elliman A, Etemad B, Fallon J, Fedirko L, Finn K, Flannery J, Forde P, Garcia-Broncano P, Gettings E, Golan D, Goodman K, Griffin A, Grimmel S, Grinke K, Hartana CA, Healy M, Heller H, Henault D, Holland G, Jiang C, Jordan H, Kaplonek P, Karlson EW, Karpell M, Kayitesi C, Lam EC, LaValle V, Lefteri K, Lian X, Lichterfeld M, Lingwood D, Liu H, Liu J, Lopez K, Lu Y, Luthern S, Ly NL, MacGowan M, Magispoc K, Marchewka J, Martino B, McNamara R, Michell A, Millstrom I, Miranda N, Nambu C, Nelson S, Noone M, Novack L, O'Callaghan C, Ommerborn C, Osborn M, Pacheco LC, Phan N, Pillai S, Porto FA, Rassadkina Y, Reissis A, Ruzicka F, Seiger K, Selleck K, Sessa L, Sharpe A, Sharr C, Shin S, Singh N, Slaughenhaupt S, Sheppard KS, Sun W, Sun X, Suschana E, Talabi O, Ticheli H, Weiss ST, Wilson V, Zhu A, The Massachusetts Consortium for Pathogen, R., SARS-CoV-2 viral load is associated with increased disease severity and mortality. Nat Commun 11(1):5493

Falzone L, Gattuso G, Tsatsakis A, Spandidos DA, Libra M, Falzone L, Gattuso G, Tsatsakis A, Spandidos DA, Libra M, Falzone L, Gattuso G, Tsatsakis A, Spandidos DA, Libra M, Falzone L, Gattuso G, Tsatsakis A, Spandidos DA, Libra M (2021) Current and innovative methods for the diagnosis of COVID-19 infection (review). Int J Mol Med 47(6):100

Fani M, Zandi M, Ebrahimi S, Soltani S, Abbasi S (2021) The role of miRNAs in COVID-19 disease. 16(4):301–306

FDA, U. S. SARS-CoV-2 viral mutations: impact on COVID-19 Tests. https://www.fda.gov/medical-devices/coronavirus-covid-19-and-medical-devices/sars-cov-2-viral-mutations-impact-covid-19-tests. Accessed 14 Feb 2022

FDA, U. S. EUA authorized serology test performance. https://www.fda.gov/medical-devices/coronavirus-disease-2019-covid-19-emergency-use-authorizations-medical-devices/eua-authorized-serology-test-performance. Accessed 14 Feb 2022

FDA, U. S

Feng W, Peng H, Xu J, Liu Y, Pabbaraju K, Tipples G, Joyce MA, Saffran HA, Tyrrell DL, Babiuk S, Zhang H, Le XC (2021) Integrating Reverse transcription recombinase polymerase amplification with CRISPR technology for the one-tube assay of RNA. Anal Chem 93(37):12808–12816

Ferré VM, Peiffer-Smadja N, Visseaux B, Descamps D, Ghosn J, Charpentier C (2022) Omicron SARS-CoV-2 variant: what we know and what we don't. Anaes Crit Care Pain Med 41(1):100998

Florea A, Melinte G, Simon I, Cristea C (2019) Electrochemical biosensors as potential diagnostic devices for autoimmune diseases. Biosensors 9(1):38

Fozouni P, Son S, Díaz de León Derby M, Knott GJ, Gray CN, D'Ambrosio MV, Zhao C, Switz NA, Kumar GR, Stephens SI, Boehm D, Tsou CL, Shu J, Bhuiya A, Armstrong M, Harris AR, Chen PY, Osterloh JM, Meyer-Franke A, Joehnk B, Walcott K, Sil A, Langelier C, Pollard KS, Crawford ED, Puschnik AS, Phelps M, Kistler A, DeRisi JL, Doudna JA, Fletcher DA, Ott M (2021) Amplification-free detection of SARS-CoV-2 with CRISPR-Cas13a and mobile phone microscopy. Cell 184(2):323–333.e9

Ganbaatar U, Liu C (2021) CRISPR-based COVID-19 testing: toward next-generation point-of-care diagnostics. Front Cell Infect Microbiol 11(373). https://creativecommons.org/licenses/by/4.0/

García-Bernalt Diego J, Fernández-Soto P, Domínguez-Gil M, Belhassen-García M, Bellido JLM, Muro A (2021) A simple, affordable, rapid, stabilized, colorimetric, versatile RT-LAMP assay to detect SARS-CoV-2. Diagnostics 11(3):438

Giovannini G, Haick H, Garoli D (2021) Detecting COVID-19 from breath: a game changer for a big challenge. ACS Sens 6(4):1408–1417

Gong F, Wei H-x, Li Q, Liu L, Li B (2021) Evaluation and comparison of serological methods for COVID-19 diagnosis. Front Mol Biosci 8

González-Donapetry P, García-Clemente P, Bloise I, García-Sánchez C, Sánchez Castellano M, Romero M. P, Gutiérrez Arroyo A, Mingorance J, de Ceano-Vivas La Calle M, García-Rodriguez J (2021) Think of the children: evaluation of SARS-CoV-2 rapid antigen test in pediatric population. Pediatr Infect Disease J 40(5):385–388

Green DA, Zucker J, Westblade LF, Whittier S, Rennert H, Velu P, Craney A, Cushing M, Liu D, Sobieszczyk ME, Boehme AK, Sepulveda JL, McAdam AJ (2020) Clinical performance of SARS-CoV-2 molecular tests. J Clin Microbiol 58(8):e00995-e1020

Grieshaber D, MacKenzie R, Vörös J, Reimhult E (2008) Electrochemical biosensors—Sensor principles and architectures. Sensors 8(3):1400–1458

Gupta A, Khurana S, Das R, Srigyan D, Singh A, Mittal A, Singh P, Soneja M, Kumar A, Singh AK, Soni KD, Meena S, Aggarwal R, Sharad N, Aggarwal A, Kadnur H, George N, Singh K, Desai D, Trilangi P, Khan AR, Kiro VV, Naik S, Arunan B, Goel S, Patidar D, Lathwal A, Dar L, Trikha A, Pandey RM, Malhotra R, Guleria R, Mathur P.; Wig N (2020) Rapid chromatographic immunoassay-based evaluation of COVID-19: A cross-sectional, diagnostic test accuracy study & its implications for COVID-19 management in India. Ind J Med Res

Habli Z, Saleh S, Zaraket H, Khraiche ML (2021) COVID-19 in-vitro diagnostics: state-of-the-art and challenges for rapid, scalable, and high-accuracy screening. Front Bioeng Biotechnol 8

Haecker H, Redecke V, Tawaratsumida K, Larragoite E, Williams E, Planelles V, Spivak A, Hirayama L, Elgort M, Swenson S, Smith R, Worthen B, Zimmerman R, Slev P, Cahoon B, Astill M (2021) A rapid and affordable point-of-care test for detection of SARS-Cov-2-specific antibodies based on hemagglutination and artificial intelligence-based image interpretation. Res Square

Hayer J, Kasapic D, Zemmrich C (2021) Real-world clinical performance of commercial SARS-CoV-2 rapid antigen tests in suspected COVID-19: a systematic meta-analysis of available data as of November 20, 2020. Int J Infect Dis 108:592–602

He Y, Xie T, Tong Y (2021a) Rapid and highly sensitive one-tube colorimetric RT-LAMP assay for visual detection of SARS-CoV-2 RNA. Biosens Bioelectron 187:113330

He Y, Wang L, An X, Tong Y (2021b) All-in-one in situ colorimetric RT-LAMP assay for point-of-care testing of SARS-CoV-2. Analyst 146(19):6026–6034

Hodge CD, Rosenberg DJ, Grob P, Wilamowski M, Joachimiak A, Hura GL, Hammel M (2021) Rigid monoclonal antibodies improve detection of SARS-CoV-2 nucleocapsid protein. mAbs 13(1):1905978

Holzner C, Pabst D, Anastasiou OE, Dittmer U, Manegold RK, Risse J, Fistera D, Kill C, Falk M (2021) SARS-CoV-2 rapid antigen test: fast-safe or dangerous? An analysis in the emergency department of an university hospital. J Med Virol 93(9):5323–5327

Hosu O, Selvolini G, Cristea C, Marrazza G (2018) Electrochemical immunosensors for disease detection and diagnosis. Curr Med Chem 25(33):4119–4137

Hu B, Guo H, Zhou P, Shi Z-L (2021) Characteristics of SARS-CoV-2 and COVID-19. Nat Rev Microbiol 19(3):141–154

Huang Y, Yang C, Xu X-F, Xu W, Liu S-W (2020a) Structural and functional properties of SARS-CoV-2 spike protein: potential antivirus drug development for COVID-19. Acta Pharmacol Sin 41(9):1141–1149

Huang Z, Tian D, Liu Y, Lin Z, Lyon CJ, Lai W, Fusco D, Drouin A, Yin X, Hu T, Ning B (2020b) Ultra-sensitive and high-throughput CRISPR-powered COVID-19 diagnosis. Biosens Bioelectron 164:112316

Huang D, Shi Z, Qian J, Bi K, Fang M, Xu Z (2021a) A CRISPR-Cas12a-derived biosensor enabling portable personal glucose meter readout for quantitative detection of SARS-CoV-2. Biotechnol Bioeng 118(4):1568–1577

Huang L, Tian S, Zhao W, Liu K, Ma X, Guo J (2021b) Aptamer-based lateral flow assay on-site biosensors. Biosens Bioelectron 186:113279

Hung K-F, Hung C-H, Hong C, Chen S-C, Sun Y-C, Wen J-W, Kuo C-H, Ko C-H, Cheng C-M (2021) Quantitative spectrochip-coupled lateral flow immunoassay demonstrates clinical potential for overcoming coronavirus disease 2019 pandemic screening challenges. Micromachines (basel) 12(3):321

Ibrahim EH, Ghramh HA, Kilany M (2021) Development of rapid and cost-effective top-loading device for the detection of anti-SARS-CoV-2 IgG/IgM antibodies. Sci Rep 11(1):14926

Ito K, Piantham C, Nishiura H (n/a) Relative instantaneous reproduction number of Omicron SARS-CoV-2 variant with respect to the Delta variant in Denmark. J Med Virol

Iyer AS, Jones FK, Nodoushani A, Kelly M, Becker M, Slater D, Mills R, Teng E, Kamruzzaman M, Garcia-Beltran WF, Astudillo M, Yang D, Miller TE, Oliver E, Fischinger S, Atyeo C, Iafrate AJ, Calderwood SB, Lauer SA, Yu J, Li Z, Feldman J, Hauser BM, Caradonna TM, Branda JA, Turbett SE, LaRocque RC, Mellon G, Barouch DH, Schmidt AG, Azman AS, Alter G, Ryan ET, Harris JB, Charles RC (2020) Persistence and decay of human antibody responses to the receptor binding domain of SARS-CoV-2 spike protein in COVID-19 patients. Sci Immunol 5(52):eabe0367

Jian MJ, Chung HY, Chang CK, Lin JC, Yeh KM, Chen CW, Li SY, Hsieh SS, Liu MT, Yang JR, Tang SH, Perng CL, Chang FY, Shang HS (2021) Clinical comparison of three sample-to-answer systems for detecting SARS-CoV-2 in B.1.1.7 lineage emergence. Infection Drug Resistance 14:3255–3261

Johns Hopkins University & Medicine Coronavirus Resource Center. https://coronavirus.jhu.edu/. Accessed 10 Feb 2022

Joung J, Ladha A, Saito M, Segel M, Bruneau R, Huang M-lW, Kim N-G, Yu X, Li J, Walker BD, Greninger AL, Jerome KR, Gootenberg JS, Abudayyeh OO, Zhang F (2020) Point-of-care testing for COVID-19 using SHERLOCK diagnostics. medRxiv 2020.05.04.20091231

Jyotsna A, Anupam D, Pranshu P, Manodeep S, Jaya G (2021) David versus Goliath: a simple antigen detection test with potential to change diagnostic strategy for SARS-CoV-2. J Infect Dev Countries 15(07)

Kabir MA, Ahmed R, Iqbal SMA, Chowdhury R, Paulmurugan R, Demirci U, Asghar W (2021) Diagnosis for COVID-19: current status and future prospects. Expert Rev Mol Diagn 21(3):269–288

Kahn M, Schuierer L, Bartenschlager C, Zellmer S, Frey R, Freitag M, Dhillon C, Heier M, Ebigbo A, Denzel C, Temizel S, Messmann H, Wehler M, Hoffmann R, Kling E, Römmele C (2021) Performance of antigen testing for diagnosis of COVID-19: a direct comparison of a lateral flow device to nucleic acid amplification based tests. BMC Infect Dis 21(1):798

Kanaujia R, Ghosh A, Mohindra R, Singla V, Goyal K, Gudisa R, Sharma V, Mohan L, Kaur N, Mohi GK, Bora I, Ratho RK, Soni RK, Bhalla A, Singh MP (2021) Rapid antigen detection kit for the diagnosis of SARS-CoV-2—Are we missing asymptomatic patients? Ind J Med Microbiol 39(4):457–461

Ke Z, Oton J, Qu K, Cortese M, Zila V, McKeane L, Nakane T, Zivanov J, Neufeldt CJ, Cerikan B, Lu JM, Peukes J, Xiong X, Kräusslich H-G, Scheres SHW, Bartenschlager R, Briggs JAG (2020) Structures and distributions of SARS-CoV-2 spike proteins on intact virions. Nature 588(7838):498–502

Kenyeres B, Ánosi N, Bányai K, Mátyus M, Orosz L, Kiss A, Kele B, Burián K, Lengyel G (2021) Comparison of four PCR and two point of care assays used in the laboratory detection of SARS-CoV-2. J Virol Methods 293:114165

Kevadiya BD, Machhi J, Herskovitz J, Oleynikov MD, Blomberg WR, Bajwa N, Soni D, Das S, Hasan M, Patel M, Senan AM, Gorantla S, McMillan J, Edagwa B, Eisenberg R, Gurumurthy CB, Reid SPM, Punyadeera C, Chang L, Gendelman HE (2021) Diagnostics for SARS-CoV-2 infections. Nat Mater 20(5):593–605

Kilic A, Hiestand B, Palavecino E, Miller MB (2021) Evaluation of performance of the BD veritor SARS-CoV-2 chromatographic immunoassay test in patients with symptoms of COVID-19. J Clin Microbiol 59(5):e00260-e321

Kim H, Hong H, Yoon SH (2020a) Diagnostic performance of CT and reverse transcriptase polymerase chain reaction for coronavirus disease 2019: a meta-analysis. Radiology 296(3):E145–E155

Kim JM, Kim HM, Lee EJ, Jo HJ, Yoon Y, Lee NJ, Son J, Lee YJ, Kim MS, Lee YP, Chae SJ, Park KR, Cho SR, Park S, Kim SJ, Wang E, Woo S, Lim A, Park SJ, Jang J, Chung YS, Chin BS, Lee JS, Lim D, Han MG, Yoo CK (2020b) Detection and isolation of SARS-CoV-2 in serum, urine, and stool specimens of COVID-19 patients from the Republic of Korea. Osong Public Health Res Perspect 11(3):112–117

Kim S, Hao Y, Miller EA, Tay DMY, Yee E, Kongsuphol P, Jia H, McBee M, Preiser PR, Sikes HD (2021) Vertical flow cellulose-based assays for SARS-CoV-2 antibody detection in human serum. ACS Sens 6(5):1891–1898

Kitajima H, Tamura Y, Yoshida H, Kinoshita H, Katsuta H, Matsui C, Matsushita A, Arai T, Hashimoto S, Iuchi A, Hirashima T, Morishita H, Matsuoka H, Tanaka T, Nagai T (2021) Clinical COVID-19 diagnostic methods: comparison of reverse transcription loop-mediated isothermal amplification (RT-LAMP) and quantitative RT-PCR (qRT-PCR). J Clin Virol: off Publ Pan Am Soc Clin Virol 139:104813

Koeleman JGM, Brand H, de Man SJ, Ong DSY (2021) Clinical evaluation of rapid point-of-care antigen tests for diagnosis of SARS-CoV-2 infection. Eur J Clin Microbiol Infect Dis 40(9):1975–1981

Kolwijck E, Brouwers-Boers M, Broertjes J, van Heeswijk K, Runderkamp N, Meijer A, Hermans MHA, Leenders ACAP (2021) Validation and implementation of the Panbio COVID-19 Ag rapid test for the diagnosis of SARS-CoV-2 infection in symptomatic hospital healthcare workers. Infect Prevent Practice 3(2):100142

Korber B, Fischer WM, Gnanakaran S, Yoon H, Theiler J, Abfalterer W, Hengartner N, Giorgi EE, Bhattacharya T, Foley B, Hastie KM, Parker MD, Partridge DG, Evans CM, Freeman TM, de Silva TI, Sheffield C-GG, McDanal C, Perez LG, Tang H, Moon-Walker A, Whelan SP, LaBranche CC, Saphire EO, Montefiori DC (2020) Tracking changes in SARS-CoV-2 spike: evidence that d614g increases infectivity of the COVID-19 virus. Cell 182(4):812-827.e19

Kostyusheva A, Brezgin S, Babin Y, Vasilyeva I, Glebe D, Kostyushev D, Chulanov V (2021) CRISPR-Cas systems for diagnosing infectious diseases. Methods

Krishnan S, Dusane A, Morajkar R, Venkat A, Vernekar AA (2021) Deciphering the role of nanostructured materials in the point-of-care diagnostics for COVID-19: a comprehensive review. J Mater Chem B 9(30):5967–5981

Krüger LJ, Gaeddert M, Tobian F, Lainati F, Gottschalk C, Klein JAF, Schnitzler P, Kräusslich HG, Nikolai O, Lindner AK, Mockenhaupt FP, Seybold J, Corman VM, Drosten C, Pollock NR, Knorr B, Welker A, de Vos M, Sacks JA, Denkinger CM (2021b) The Abbott PanBio WHO emergency use listed, rapid, antigen-detecting point-of-care diagnostic test for SARS-CoV-2-evaluation of the accuracy and ease-of-use. PLoS ONE 16(5):e0247918

Krüger LJ, Klein JAF, Tobian F, Gaeddert M, Lainati F, Klemm S, Schnitzler P, Bartenschlager R, Cerikan B, Neufeldt CJ, Nikolai O, Lindner AK, Mockenhaupt FP, Seybold J, Jones TC, Corman VM, Pollock NR, Knorr B, Welker A, de Vos M, Sacks JA, Denkinger CM (2021a) Evaluation of accuracy, exclusivity, limit-of-detection and ease-of-use of LumiraDx™: an antigen-detecting point-of-care device for SARS-CoV-2. Infection1–12

Kruse RL, Huang Y, Smetana H, Gehrie EA, Amukele TK, Tobian AAR, Mostafa HH, Wang ZZ (2021) A rapid, point-of-care red blood cell agglutination assay detecting antibodies against SARS-CoV-2. Biochem Biophys Res Commun 553:165–171

Kruse RL, Huang Y, Lee A, Zhu X, Shrestha R, Laeyendecker O, Littlefield K, Pekosz A, Bloch EM, Tobian AAR, Wang ZZ (2021). A hemagglutination-based, semi-quantitative test for point-of-care determination of SARS-CoV-2 antibody levels. J Clin Microbiol 0(ja):JCM.01186-21

Kudr J, Michalek P, Ilieva L, Adam V, Zitka O (2021) COVID-19: a challenge for electrochemical biosensors. TrAC Trends Anal Chem 136:116192

Lai CKC, Lam W (2021) Laboratory testing for the diagnosis of COVID-19. Biochem Biophys Res Commun 538:226–230

Landaas ET, Storm ML, Tollånes MC, Barlinn R, Kran AB, Bragstad K, Christensen A, Andreassen T (2021) Diagnostic performance of a SARS-CoV-2 rapid antigen test in a large, Norwegian cohort. J Clin Virol: off Publ Pan Am Soc Clin Virol 137:104789

Lauring AS, Hodcroft EB (2021) Genetic variants of SARS-CoV-2—What do they mean? JAMA 325(6):529–531

Lee J, Song J-U (2021) Diagnostic accuracy of the Cepheid Xpert Xpress and the Abbott ID NOW assay for rapid detection of SARS-CoV-2: a systematic review and meta-analysis. J Med Virol 93(7):4523–4531

Lee D, Chu C-H, Sarioglu AF (2021) Point-of-care toolkit for multiplex molecular diagnosis of SARS-CoV-2 and influenza A and B viruses. ACS Sens 6(9):3204–3213

Li J, Lillehoj PB (2021) Microfluidic magneto immunosensor for rapid, high sensitivity measurements of SARS-CoV-2 nucleocapsid protein in serum. ACS Sens 6(3):1270–1278

Li J, Lin R, Yang Y, Zhao R, Song S, Zhou Y, Shi J, Wang L, Song H, Hao R (2021) Multichannel immunosensor platform for the rapid detection of SARS-CoV-2 and influenza A(H1N1) virus. ACS Appl Mater Interfaces 13(19):22262–22270

Lisboa Bastos M, Tavaziva G, Abidi SK, Campbell JR, Haraoui L-P, Johnston JC, Lan Z, Law S, MacLean E, Trajman A, Menzies D, Benedetti A, Ahmad Khan F (2020) Diagnostic accuracy of serological tests for covid-19: systematic review and meta-analysis. BMJ 370:m2516

Liu Y, Rocklöv J (2021) The reproductive number of the Delta variant of SARS-CoV-2 is far higher compared to the ancestral SARS-CoV-2 virus. J Travel Med 28(7)

Luminex ARIES® SARS-CoV-2 assay. https://www.luminexcorp.com/aries-sars-cov-2-assay/#eua. Accessed 14 Feb 2022

Mahas A, Wang Q, Marsic T, Mahfouz MM (2021) A novel miniature CRISPR-Cas13 system for SARS-CoV-2 diagnostics. ACS Synth Biol 10(10):2541–2551

Mahshid SS, Flynn SE, Mahshid S (2021) The potential application of electrochemical biosensors in the COVID-19 pandemic: A perspective on the rapid diagnostics of SARS-CoV-2. Biosens Bioelectron 176:112905

Mardian Y, Kosasih H, Karyana M, Neal, A, Lau C-Y (2021) Review of current COVID-19 diagnostics and opportunities for further development. Front Med 8(562)

Marsic T, Ali Z, Tehseen M, Mahas A, Hamdan S, Mahfouz M (2021) Vigilant: an engineered VirD2-Cas9 complex for lateral flow assay-based detection of SARS-CoV2. Nano Lett 21(8):3596–3603

Mavrikou S, Moschopoulou G, Tsekouras V, Kintzios S (2020) Development of a portable, ultra-rapid and ultra-sensitive cell-based biosensor for the direct detection of the SARS-CoV-2 S1 spike protein antigen. Sensors 20(11):3121

Mavrikou S, Tsekouras V, Hatziagapiou K, Paradeisi F, Bakakos P, Michos A, Koutsoukou A, Konstantellou E, Lambrou GI, Koniari E, Tatsi E-B, Papaparaskevas J, Iliopoulos D, Chrousos GP, Kintzios S (2021) Clinical application of the novel cell-based biosensor for the ultra-rapid detection of the SARS-CoV-2 S1 spike protein antigen: a practical approach. Biosensors 11(7):224

Mboumba Bouassa R-S, Veyer D, Péré H, Bélec L (2021) Analytical performances of the point-of-care SIENNA™ COVID-19 antigen rapid test for the detection of SARS-CoV-2 nucleocapsid protein in nasopharyngeal swabs: a prospective evaluation during the COVID-19 second wave in France. Int J Infect Dis 106:8–12

NS Medical Devices MediCircle Health launches SpectraLIT Covid-19 test in India. https://www.nsmedicaldevices.com/news/medicircle-health-spectralit-covid-19-india/. Accessed 14 Feb 2022

Mekaliah CR, Farag, D'Andra G, Christian H, Danielle H, Ashani M, Brandon V, Josef W, Ukamaka Diké S, Briana J, Marlon SH (2021) A review of the SalivaDirect test for COVID-19. U.S. Pharmacist pp 39–42

Mercer TR, Salit M (2021) Testing at scale during the COVID-19 pandemic. Nat Rev Genet 22(7):415–426

Michelena PD, Torres I, Ramos-García Á, Gosalbes V, Ruiz N, Sanmartín A, Botija P, Poujois S, Huntley D, Albert E, Navarro D (2022) Real-life performance of a COVID-19 rapid antigen detection test targeting the SARS-CoV-2 nucleoprotein for diagnosis of COVID-19 due to the Omicron variant. medRxiv 2022.02.02.22270295

Mina MJ, Andersen KG (2021) COVID-19 testing: One size does not fit all. Science 371(6525):126–127

Mittal A, Gupta A, Kumar S, Surjit M, Singh B, Soneja M, Soni KD, Khan AR, Singh K, Naik S, Kumar A, Aggarwal R, Nischal N, Sinha S, Trikha A, Wig N (2020) Gargle lavage as a viable alternative to swab for detection of SARS-CoV-2. Ind J Med Res 152(1 & 2):77–81

Monošík R, Streďanský M, Šturdík E (2012) Application of electrochemical biosensors in clinical diagnosis. J Clin Lab Anal 26(1):22–34

Mor M, Waisman Y (2000) Point-of-care testing: a critical review. Pediatr Emerg Care 16(1):45–48

Muhi S, Tayler N, Hoang T, Ballard SA, Graham M, Rojek A, Kwong JC, Trubiano JA, Smibert O, Drewett G, James F, Gardiner E, Chea S, Isles N, Sait M, Pasricha S, Taiaroa G, McAuley J, Williams E, Gibney KB, Stinear TP, Bond K, Lewin SR, Putland M, Howden BP, Williamson DA (2021) Multi-site assessment of rapid, point-of-care antigen testing for the diagnosis of SARS-CoV-2 infection in a low-prevalence setting: a validation and implementation study. Lancet Region Health Western Pac 9:100115

Munne K, Bhanothu V, Bhor V, Patel V, Mahale SD, Pande S (2021) Detection of SARS-CoV-2 infection by RT-PCR test: factors influencing interpretation of results. VirusDisease 32(2):187–189

Ng D, Pinharanda A, Vogt MC, Litwin-Kumar A, Stearns K, Thopte U, Cannavo E, Enikolopov A, Fiederling F, Kosmidis S, Noro B, Rodrigues-Vaz I, Shayya H, Andolfatto P, Peterka DS, Tabachnik T, D'Armiento J, Goldklang M, Bendesky A (2021) WHotLAMP: a simple, inexpensive, and sensitive molecular test for the detection of SARS-CoV-2 in saliva. PLoS ONE 16(9):e0257464–e0257464

Nisa F, Raspati CK, Basti A, Lia F, Silvita FR, Leonardus W, Delita P, Savira E, Azzania F, Emma R, Ryan BR, Cut NCA, Susantina P, Bony WL, Ida PS (2021) The performance of point-of-care antibody test for COVID-19 diagnosis in a tertiary hospital in Bandung, Indonesia. J Infect Dev Countries 15(02)

Nordgren J, Sharma S, Olsson H, Jämtberg M, Falkeborn T, Svensson L, Hagbom M (2021) SARS-CoV-2 rapid antigen test: high sensitivity to detect infectious virus. J Clin Virol: off Publ Pan Am Soc Clin Virol 140:104846

Nouri R, Tang Z, Dong M, Liu T, Kshirsagar A, Guan W (2021) CRISPR-based detection of SARS-CoV-2: a review from sample to result. Biosens Bioelectron 178:113012

Obande GA, Banga Singh KK (2020) Current and future perspectives on isothermal nucleic acid amplification technologies for diagnosing infections. Infect Drug Resist 13:455–483

Olearo F, Nörz D, Heinrich F, Sutter JP, Roedl K, Schultze A, Wiesch JSZ, Braun P, Oestereich L, Kreuels B, Wichmann D, Aepfelbacher M, Pfefferle S, Lütgehetmann M (2021) Handling and accuracy of four rapid antigen tests for the diagnosis of SARS-CoV-2 compared to RT-qPCR. J Clin Virol: off Publ Pan Am Soc Clin Virol 137:104782

Abbott evaluating Omicron and other COVID variants to ensure test effectiveness. https://www.abbott.com/corpnewsroom/diagnostics-testing/monitoring-covid-variants-to-ensure-test-effectiveness.html. Accessed 4 Feb 2022

Osterman A, Iglhaut M, Lehner A, Späth P, Stern M, Autenrieth H, Muenchhoff M, Graf A, Krebs S, Blum H, Baiker A, Grzimek-Koschewa N, Protzer U, Kaderali L, Baldauf HM, Keppler OT (2021) Comparison of four commercial, automated antigen tests to detect SARS-CoV-2 variants of concern. Med Microbiol Immunol 210(5–6):263–275

Pallett SJC, Denny SJ, Patel A, Charani E, Mughal N, Stebbing J, Davies GW, Moore LSP (2021) Point-of-care SARS-CoV-2 serological assays for enhanced case finding in a UK inpatient population. Sci Rep 11(1):5860

Panferov VG, Byzova NA, Biketov SF, Zherdev AV, Dzantiev BB (2021) Comparative study of in situ techniques to enlarge gold nanoparticles for highly sensitive lateral flow immunoassay of SARS-CoV-2. Biosensors 11(7):229

Pascarella G, Strumia A, Piliego C, Bruno F, Del Buono R, Costa F, Scarlata S, Agrò FE (2020) COVID-19 diagnosis and management: a comprehensive review. J Int Med 288(2):192–206

Paton TF, Marr I, O'Keefe Z, Inglis TJJ (2021) Development, deployment and in-field demonstration of mobile coronavirus SARS-CoV-2 nucleic acid amplification test. J Med Microbiol 70(4)

Perry BL, Aronson B, Railey AF, Ludema C (2021) If you build it, will they come? Social, economic, and psychological determinants of COVID-19 testing decisions. PLoS ONE 16(7):e0252658

Pietschmann J, Voepel N, Voß L, Rasche S, Schubert M, Kleines M, Krause H-J, Shaw TM, Spiegel H, Schroeper F (2021) Development of fast and portable frequency magnetic mixing-based serological SARS-CoV-2-specific antibody detection assay. Front Microbiol 12(841)

Pilarowski G, Marquez C, Rubio L, Peng J, Martinez J, Black D, Chamie G, Jones D, Jacobo J, Tulier-Laiwa V, Rojas S, Rojas S, Cox C, Nakamura R, Petersen M, DeRisi J, Havlir DV (2020) Field performance and public health response using the BinaxNOWTM rapid severe acute respiratory syndrome coronavirus 2 (SARS-CoV-2) antigen detection assay during community-based testing. Clin Infect Dis 73(9):e3098–e3101

Pinto B, Brito D, Valadão D, Alves D, Heimfarth L (2020) Can the Wondfo® SARS-CoV-2 IgM/IgG antibodies be used as a rapid diagnostic test? Arch Biotechnol Biomed 4:013–017

Pollock NR, Jacobs JR, Tran K, Cranston AE, Smith S, O'Kane CY, Roady TJ, Moran A, Scarry A, Carroll M, Volinsky L, Perez G, Patel P, Gabriel S, Lennon NJ, Madoff LC, Brown C, Smole SC, Loeffelholz MJ (2021) Performance and implementation evaluation of the Abbott BinaxNOW Rapid antigen test in a high-throughput drive-through community testing site in Massachusetts. J Clin Microbiol 59(5):e00083-e121

Prince-Guerra JL, Almendares O, Nolen LD, et al (2021) Evaluation of Abbott BinaxNOW rapid antigen test for SARS-CoV-2 infection at two community-based testing sites—Pima County, Arizona, 3–17 Nov 2020. MMWR Morb Mortal Wkly Rep 70:100–105

Rahman MR, Hossain MA, Mozibullah M, Mujib FA, Afrose A, Shahed-Al-Mahmud M, Apu MAI (2021) CRISPR is a useful biological tool for detecting nucleic acid of SARS-CoV-2 in human clinical samples. Biomed Pharmacother 140:111772

Rahmati Z, Roushani M, Hosseini H, Choobin H (2021) Electrochemical immunosensor with Cu_2O nanocube coating for detection of SARS-CoV-2 spike protein. Microchim Acta 188(3):105

Razo SC, Panferov VG, Safenkova IV, Varitsev YA, Zherdev AV, Pakina EN, Dzantiev BB (2018) How to improve sensitivity of sandwich lateral flow immunoassay for corpuscular antigens on the example of potato virus Y? Sensors 18(11):3975

Regunath H (2022) COVID-19 pills—A long awaited ally for out-patient therapeutics. Missouri Med 27

Rochman ND, Wolf YI, Faure G, Mutz P, Zhang F, Koonin EV (2021) Ongoing global and regional adaptive evolution of SARS-CoV-2. Proc Natl Acad Sci 118(29):e2104241118

Rodgers MA, Olivo A, Harris BJ, Lark C, Luo X, Berg MG, Meyer TV, Mohaimani A, Orf GS, Goldstein Y, Fox AS, Hirschhorn J, Glen WB, Nolte F, Landay A, Jennings C, Moy J, Servellita V, Chiu C, Batra R, Snell LB, Nebbia G, Douthwaite S, Tanuri A, Singh L, de Oliveira T, Ahouidi A, Mboup S, Cloherty GA (2022) Detection of SARS-CoV-2 variants by Abbott molecular, antigen, and serological tests. J Clin Virol 147:105080

Rodriguez-Manzano J, Malpartida-Cardenas K, Moser N, Pennisi I, Cavuto M, Miglietta L, Moniri A, Penn R, Satta G, Randell P, Davies F, Bolt F, Barclay W, Holmes A, Georgiou P (2021) Handheld point-of-care system for rapid detection of SARS-CoV-2 extracted RNA in under 20 min. ACS Cent Sci 7(2):307–317

Rodriguez-Mateos P, Ngamsom B, Walter C, Dyer CE, Gitaka J, Iles A, Pamme N (2021) A lab-on-a-chip platform for integrated extraction and detection of SARS-CoV-2 RNA in resource-limited settings. Anal Chim Acta 1177:338758

Russo A, Minichini C, Starace M, Astorri R, Calò F, Coppola N (2020) Current status of laboratory diagnosis for COVID-19: a narrative review. Infect Drug Resist 13:2657–2665

Salvagno GL, Nocini R, Gianfilippi G, Fiorio G, Pighi L, Nitto SD, Cominziolli A, Henry BM, Lippi G (2022) Performance of Fujirebio Espline SARS-CoV-2 rapid antigen test for identifying potentially infectious individuals. Diagnosis 9(1):146–148

Samprathi M, Jayashree M (2021) Biomarkers in COVID-19: an up-to-date review. Front Pediat 8. https://creativecommons.org/licenses/by-nc/4.0/

Schellenberg JJ, Ormond M, Keynan Y (2021) Extraction-free RT-LAMP to detect SARS-CoV-2 is less sensitive but highly specific compared to standard RT-PCR in 101 samples. J Clin Virol 136:104764

Schuler CFT, Gherasim C, O'Shea K, Manthei DM, Chen J, Giacherio D, Troost JP, Baldwin JL, Baker JR Jr (2021) Accurate point-of-care serology tests for COVID-19. PLoS One 16(3):e0248729

Seynaeve Y, Heylen J, Fontaine C, Maclot F, Meex C, Diep AN, Donneau A-F, Hayette M-P, Descy J (2021) Evaluation of two rapid antigenic tests for the detection of SARS-CoV-2 in nasopharyngeal swabs. J Clin Med 10(13)

Shaw JLV, Deslandes V, Smith J, Desjardins M (2021) Evaluation of the Abbott PanbioTM COVID-19 Ag rapid antigen test for the detection of SARS-CoV-2 in asymptomatic Canadians. Diagn Microbiol Infect Dis 101(4):115514

Shyu D, Dorroh J, Holtmeyer C, Ritter D, Upendran A, Kannan R, Dandachi D, Rojas-Moreno C, Whitt SP, Regunath H (2020) Laboratory tests for COVID-19: a review of peer-reviewed publications and implications for clinical use. Missouri Med 117(3):184–195

Silveira MF, Mesenburg MA, Dellagostin OA, de Oliveira NR, Maia MA, Santos FD, Vale A, Menezes AMB, Victora GD, Victora CG, Barros AJ, Vidaletti LP, Hartwig FP, Barros FC, Hallal PC, Horta BL (2021) Time-dependent decay of detectable antibodies against SARS-CoV-2: A comparison of ELISA with two batches of a lateral-flow test. Braz J Infect Dis: off Publ Braz Soc Infect Dis 25(4):101601

Singh NK, Ray P, Carlin AF, Magallanes C, Morgan SC, Laurent LC, Aronoff-Spencer ES, Hall DA (2021) Hitting the diagnostic sweet spot: point-of-care SARS-CoV-2 salivary antigen testing with an off-the-shelf glucometer. Biosens Bioelectron 180:113111–113111

Song Q, Sun X, Dai Z, Gao Y, Gong X, Zhou B, Wu J, Wen W (2021a) Point-of-care testing detection methods for COVID-19. Lab Chip 21(9):1634–1660

Song J, El-Tholoth M, Li Y, Graham-Wooten J, Liang Y, Li J, Li W, Weiss SR, Collman RG, Bau HH (2021b) Single- and two-stage, closed-tube, point-of-care, molecular detection of SARS-CoV-2. Anal Chem 93(38):13063–13071

Soroka M, Wasowicz B, Rymaszewska A (2021) Loop-mediated isothermal amplification (LAMP): the better sibling of PCR? Cells 10(8):1931. https://creativecommons.org/licenses/by/4.0/

Soung YH, Ford S, Zhang V, Chung J (2017) Exosomes Cancer Diagnost 9(1):8

Stambaugh A, Parks JW, Stott MA, Meena GG, Hawkins AR, Schmidt H (2021) Optofluidic multiplex detection of single SARS-CoV-2 and influenza A antigens using a novel bright fluorescent probe assay. Proc Natl Acad Sci 118(20):e2103480118

Strand R, Thelaus L, Fernström N, Sunnerhagen T, Lindroth Y, Linder A, Rasmussen M (2021) Rapid diagnostic testing for SARS-CoV-2: validation and comparison of three point-of-care antibody tests. J Med Virol 93(7):4592–4596

Sun Y, Yu L, Liu C, Ye S, Chen W, Li D, Huang W (2021) One-tube SARS-CoV-2 detection platform based on RT-RPA and CRISPR/Cas12a. J Transl Med 19(1):74

Sundah NR, Natalia A, Liu Y, Ho NRY, Zhao H, Chen Y, Miow QH, Wang Y, Beh DLL, Chew KL, Chan D, Tambyah PA, Ong CWM, Shao H (2021) Catalytic amplification by transition-state molecular switches for direct and sensitive detection of SARS-CoV-2. Sci Adv 7(12):eabe5940

Svobodova M, Skouridou V, Jauset-Rubio M, Viéitez I, Fernández-Villar A, Cabrera Alvargonzalez JJ, Poveda E, Bofill CB, Sans T, Bashammakh A, Alyoubi AO, O'Sullivan CK (2021) Aptamer sandwich assay for the detection of SARS-CoV-2 spike protein antigen. ACS Omega 6(51):35657–35666

Szunerits, S.; Pagneux, Q.; Swaidan, A.; Mishyn, V.; Roussel, A.; Cambillau, C.; Devos, D.; Engelmann, I.; Alidjinou, E. K.; Happy, H.; Boukherroub, R., The role of the surface ligand on the performance of electrochemical SARS-CoV-2 antigen biosensors. Anal Bioanal Chem

Szymczak WA, Goldstein DY, Orner EP, Fecher RA, Yokoda RT, Skalina KA, Narlieva M, Gendlina I, Fox AS (2020) Utility of stool PCR for the diagnosis of COVID-19: comparison of two commercial platforms. J Clin Microbiol 58(9):e01369-e1420

Taleghani N, Taghipour F (2021) Diagnosis of COVID-19 for controlling the pandemic: a review of the state-of-the-art. Biosens Bioelectron 174:112830

Teymouri M, Mollazadeh S, Mortazavi H, Naderi Ghale-noie Z, Keyvani V, Aghababaei F, Hamblin MR, Abbaszadeh-Goudarzi G, Pourghadamyari H, Hashemian SMR, Mirzaei H (2021) Recent advances and challenges of RT-PCR tests for the diagnosis of COVID-19. Pathol Res Practice 221:153443

Tinker S, Szablewski C, Litvintseva A, Drenzek C, Voccio G, Hunter M, Briggs S, Heida D, Folster J, Shewmaker P, Medrzycki M, Bowen M, Bohannon C, Bagarozzi D, Petway M, Rota P, Kuhnert-Tallman W, Thornburg N, Prince-Guerra J, Barrios L, Tamin A, Harcourt J, Honein M (2021) Point-of-care antigen test for SARS-CoV-2 in asymptomatic college students. Emerging Infect Disease J 27(10):2662

Tollånes MC, Jenum PA, Kierkegaard H, Abildsnes E, Bævre-Jensen RM, Breivik AC, Sandberg S (2021) Evaluation of 32 rapid tests for detection of antibodies against SARS-CoV-2. Clin Chim Acta 519:133–139

Torres I, Poujois S, Albert E, Álvarez G, Colomina J, Navarro D (2021a) Point-of-care evaluation of a rapid antigen test (CLINITEST(®) rapid COVID-19 antigen test) for diagnosis of SARS-CoV-2 infection in symptomatic and asymptomatic individuals. J Infect 82(5):e11–e12

Torres MDT, de Araujo WR, de Lima LF, Ferreira AL, de la Fuente-Nunez C (2021b) Low-cost biosensor for rapid detection of SARS-CoV-2 at the point of care. Matter 4(7):2403–2416

Townsend A, Rijal P, Xiao J, Tan TK, Huang K-YA, Schimanski L, Huo J, Gupta N, Rahikainen R, Matthews PC, Crook D, Hoosdally S, Dunachie S, Barnes E, Street T, Conlon CP, Frater J, Arancibia-Cárcamo CV, Rudkin J, Stoesser N, Karpe F, Neville M, Ploeg R, Oliveira M, Roberts DJ, Lamikanra AA, Tsang HP, Bown A, Vipond R, Mentzer AJ, Knight JC, Kwok AJ, Screaton GR, Mongkolsapaya J, Dejnirattisai W, Supasa P, Klenerman P, Dold C, Baillie JK, Moore SC, Openshaw PJM, Semple MG, Turtle LCW, Ainsworth M, Allcock A, Beer S, Bibi S, Skelly D, Stafford L, Jeffrey K, O'Donnell D, Clutterbuck E, Espinosa A, Mendoza M, Georgiou D, Lockett T, Martinez J, Perez E, Gallardo Sanchez V, Scozzafava G, Sobrinodiaz A, Thraves H, Joly E (2021) A haemagglutination test for rapid detection of antibodies to SARS-CoV-2. Nat Commun 12(1):1951

Treggiari D, Piubelli C, Caldrer S, Mistretta M, Ragusa A, Orza P, Pajola B, Piccoli D, Conti A, Lorenzi C, Serafini V, Boni M, Perandin F (2022) SARS-CoV-2 rapid antigen test in comparison to RT-PCR targeting different genes: a real-life evaluation among unselected patients in a regional hospital of Italy. J Med Virol 94(3):1190–1195

Trypsteen W, Van Cleemput J, Snippenberg WV, Gerlo S, Vandekerckhove L (2020) On the whereabouts of SARS-CoV-2 in the human body: a systematic review. PLoS Pathog 16(10):e1009037

U.S. FDA. In vitro diagnostics EUAs—Molecular diagnostic tests for SARS-CoV-2. https://www.fda.gov/medical-devices/coronavirus-disease-2019-covid-19-emergency-use-authorizations-medical-devices/in-vitro-diagnostics-euas-molecular-diagnostic-tests-sars-cov-2. Accessed 3 Mar 2022a

U.S. FDA. In vitro diagnostics EUAs—Antigen diagnostic tests for SARS-CoV-2. https://www.fda.gov/medical-devices/coronavirus-disease-2019-covid-19-emergency-use-authorizations-medical-devices/in-vitro-diagnostics-euas-antigen-diagnostic-tests-sars-cov-2. Accessed 28 Feb 2022b

Ulinici M, Covantev S, Wingfield-Digby J, Beloukas A, Mathioudakis AG, Corlateanu A (2021) Screening, diagnostic and prognostic tests for COVID-19: a comprehensive review. Life 11(6):561

Valera E, Jankelow A, Lim J, Kindratenko V, Ganguli A, White K, Kumar J, Bashir R (2021) COVID-19 point-of-care diagnostics: present and future. ACS Nano 15(5):7899–7906

Valones MA, Guimarães RL, Brandão LA, de Souza PR, de Albuquerque Tavares Carvalho A, Crovela S (2009) Principles and applications of polymerase chain reaction in medical diagnostic fields: a review. Braz J Microbiol 40(1):1–11 (Publication of the Brazilian Society for Microbiology)

Vandenberg O, Martiny D, Rochas O, van Belkum A, Kozlakidis Z (2021) Considerations for diagnostic COVID-19 tests. Nat Rev Microbiol 19(3):171–183

Varona M, Eitzmann DR, Anderson JL (2021) Sequence-specific detection of ORF1a, BRAF, and ompW DNA sequences with loop mediated isothermal amplification on lateral flow immunoassay strips enabled by molecular beacons. Anal Chem 93(9):4149–4153

Vengesai A, Midzi H, Kasambala M, Mutandadzi H, Mduluza-Jokonya TL, Rusakaniko S, Mutapi F, Naicker T, Mduluza T (2021) A systematic and meta-analysis review on the diagnostic accuracy of antibodies in the serological diagnosis of COVID-19. Syst Rev 10(1):155

Wang X, Tan L, Wang X, Liu W, Lu Y, Cheng L, Sun Z (2020) Comparison of nasopharyngeal and oropharyngeal swabs for SARS-CoV-2 detection in 353 patients received tests with both specimens simultaneously. Int J Infect Dis 94:107–109

Wang Y, Zhang Y, Chen J, Wang M, Zhang T, Luo W, Li Y, Wu Y, Zeng B, Zhang K, Deng R, Li W (2021a) Detection of SARS-CoV-2 and Its mutated variants via CRISPR-Cas13-based transcription amplification. Anal Chem 93(7):3393–3402

Wang C, Cheng X, Liu L, Zhang X, Yang X, Zheng S, Rong Z, Wang S (2021b) Ultrasensitive and simultaneous detection of two specific SARS-CoV-2 antigens in human specimens using direct/enrichment dual-mode fluorescence lateral flow immunoassay. ACS Appl Mater Interfaces 13(34):40342–40353

Wen D, Yang S, Li G, Xuan Q, Guo W, Wu W (2021) Sample-to-answer and routine real-time RT-PCR: a comparison of different platforms for SARS-CoV-2 detection. J Mol Diagnos: JMD 23(6):665–670

West R, Gronvall GK, Kobokovich A (2021) Variants, vaccines and what they mean For COVID-19 testing. https://publichealth.jhu.edu/2021/variants-vaccines-and-what-they-mean-for-covid-19-testing. Accessed 14 Feb 2022

Williams E, Bond K, Zhang B, Putland M, Williamson DA, McAdam AJ (2020) Saliva as a noninvasive specimen for detection of SARS-CoV-2. J Clin Microbiol 58(8):e00776-e820

Winter AK, Hegde ST (2020) The important role of serology for COVID-19 control. Lancet Infect Dis 20(7):758–759

World Health Organization Antigen-detection in the diagnosis of SARS-CoV-2 infection. https://www.who.int/publications/i/item/antigen-detection-in-the-diagnosis-of-sars-cov-2infection-using-rapid-immunoassays. Accessed 14 Feb 2022

Wu S, Shi X, Chen Q, Jiang Y, Zuo L, Wang L, Jiang M, Lin Y, Fang S, Peng B, Wu W, Liu H, Zhang R, Kwan PSL, Hu Q (2021) Comparative evaluation of six nucleic acid amplification kits for SARS-CoV-2 RNA detection. Ann Clin Microbiol Antimicrob 20(1):38

Xu W, Liu J, Song D, Li C, Zhu A, Long F (2021) Rapid, label-free, and sensitive point-of-care testing of anti-SARS-CoV-2 IgM/IgG using all-fiber Fresnel reflection microfluidic biosensor. Mikrochim Acta 188(8):261

Xun G, Lane ST, Petrov VA, Pepa BE, Zhao H (2021) A rapid, accurate, scalable, and portable testing system for COVID-19 diagnosis. Nat Commun 12(1):2905. https://creativecommons.org/licenses/by/4.0/

Yadav R, Chaudhary JK, Jain N, Chaudhary PK, Khanra S, Dhamija P, Sharma A, Kumar A, Handu S (2021a) Role of structural and non-structural proteins and therapeutic targets of SARS-CoV-2 for COVID-19. Cells 10(4):821

Yadav AK, Verma D, Kumar A, Kumar P, Solanki PR (2021b) The perspectives of biomarker-based electrochemical immunosensors, artificial intelligence and the Internet of Medical Things toward COVID-19 diagnosis and management. Mater Today Chem 20:100443

Yamamoto S, Tanaka A, Kobayashi S, Oshiro Y, Ozeki M, Maeda K, Matsuda K, Miyo K, Mizoue T, Sugiura W, Mitsuya H, Sugiyama H, Ohmagari N (2021) Consistency of the results of rapid serological tests for SARS-CoV-2 among healthcare workers in a large national hospital in Tokyo, Japan. Glob Health Med 3(2):90–94

Yang Q, Meyerson NR, Clark SK, Paige CL, Fattor WT, Gilchrist AR, Barbachano-Guerrero A, Healy BG, Worden-Sapper ER, Wu SS, Muhlrad D, Decker CJ, Saldi TK, Lasda E, Gonzales P, Fink MR, Tat KL, Hager CR, Davis JC, Ozeroff CD, Brisson GR, McQueen MB, Leinwand LA, Parker R, Sawyer SL (2021a) Saliva TwoStep for rapid detection of asymptomatic SARS-CoV-2 carriers. Elife 10:e65113

Yang Y, Liu J, Zhou X (2021b) A CRISPR-based and post-amplification coupled SARS-CoV-2 detection with a portable evanescent wave biosensor. Biosens Bioelectron 190:113418–113418

Ye X, Wang N, Li Y, Fang X, Kong J (2021) A high-specificity flap probe-based isothermal nucleic acid amplification method based on recombinant FEN1-Bst DNA polymerase. Biosens Bioelectron 192:113503

Yüce M, Filiztekin E, Özkaya KG (2021) COVID-19 diagnosis -A review of current methods. Biosens Bioelectron 172:112752

Zander J, Scholtes S, Ottinger M, Kremer M, Kharazi A, Stadler V, Bickmann J, Zeleny C, Kuiper JWP, Hauck CR (2021) Self-collected gargle lavage allows reliable detection of SARS-CoV-2 in an outpatient setting. Microbiology Spectrum 9(1):e0036121

Zhang Z, Wang X, Wei X, Zheng SW, Lenhart BJ, Xu P, Li J, Pan J, Albrecht H, Liu C (2021a) Multiplex quantitative detection of SARS-CoV-2 specific IgG and IgM antibodies based on DNA-assisted nanopore sensing. Biosens Bioelectron 181:113134

Zhang S, Amahong K, Sun X, Lian X, Liu J, Sun H, Lou Y, Zhu F, Qiu Y (2021b) The miRNA: a small but powerful RNA for COVID-19. Brief Bioinform 22(2):1137–1149

Zhang L, Jackson CB, Mou H, Ojha A, Rangarajan E. S, Izard T, Farzan M, Choe H (2020) The D614G mutation in the SARS-CoV-2 spike protein reduces S1 shedding and increases infectivity. bioRxiv 2020.06.12.148726

Zonneveld R, Jurriaans S, van Gool T, Hofstra JJ, Hekker TAM, Defoer P, Broekhuizen-van Haaften PE, Wentink-Bonnema EM, Boonkamp L, Teunissen CE, Heijboer AC, Martens F, de Bree G, van Vugt M, van Houdt R (2021) Head-to-head validation of six immunoassays for SARS-CoV-2 in hospitalized patients. J Clin Virol: off Publ Pan Am Soc Clin Virol 139:104821

Zowawi HM, Alenazi TH, AlOmaim WS, Wazzan A, Alsufayan A, Hasanain RA, Aldibasi OS, Althawadi S, Altamimi SA, Mutabagani M, Alamri M, Almaghrabi RS, Al-Abdely HM, Memish ZA, Alqahtani SA (2021) Portable RT-PCR system: a rapid and scalable diagnostic tool for COVID-19 testing. J Clin Microbiol 59(5)

Graphical Abstract

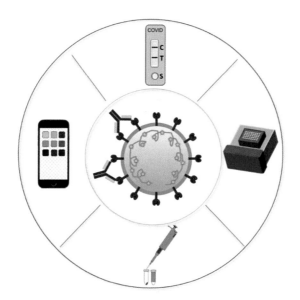

A. Gangula et al., *Point-of-Care Testing of COVID-19*,
Nanotheranostics, https://doi.org/10.1007/978-981-19-4957-9

Printed in the United States
by Baker & Taylor Publisher Services